History of Crustacea

By

Fritz Muller

tion et mise en page : Culturea (Hérault, 34)
rouvez notre catalogue sur http://culturea.fr/
tact : infos@culturea.fr
rimé en Allemagne par Books on Demand Gmbh
ign typographique et layout : Derek Murphy et Reedsy
N : 9791041985791
e de parution : février 2024
ivre :
été composé avec une police open source libre de droits dessinée sur Open Fondry https://open-foundry.com/

ermet de replanter un arbre. SHS Éditions soutient Plantons Pour l'Avenir, fonds de dotation pour le reboisement des forêts françaises
0 projets de reboisement de forêts soutenus depuis 2014 à travers la France https://www.plantonspourlavenir.fr/

CHAPTER 1. INTRODUCTORY.

When I had read Charles Darwin's book 'On the Origin of Species,' it seemed to me that there was one mode, and that perhaps the most certain, of testing the correctness of the views developed in it, namely, to attempt apply them as specially as possible to some particular group of animals. such an attempt to establish a genealogical tree, whether for the families of a class, the genera of a large family, or for the species of an extensive genus, and to produce pictures as complete and intelligible as possible of the common ancestors of the various smaller and larger circles, might furnish a result in three different ways.

1. In the first place, Darwin's suppositions when thus applied might lead to irreconcilable and contradictory conclusions, from which the erroneousness of the suppositions might be inferred. If Darwin's opinions are false, it was to be expected that contradictions would accompany their detailed application at every step, and that these, by their cumulative force, would entirely destroy the suppositions from which they proceeded, even though the deductions derived from each particular case might possess little of the unconditional nature of mathematical proof.

2. Secondly, the attempt might be successful to a greater or less extent. If it was possible upon the foundation and with the aid of the Darwinian theory, to show in what sequence the various smaller and larger circles had separated from the common fundamental form and from each other, in what sequence they had acquired the peculiarities which now characterise them, and what transformations they had undergone in the lapse of ages,—if the establishment of such a genealogical tree, of a primitive history of the group under consideration, free from internal contradictions, was possible,—then this conception, the more completely it took up all the species within itself, and the more deeply it enabled us to descend into the details of their structure, must in the same proportion bear in itself the warrant of its truth, and the more convincingly prove that the foundation upon which it is built is no loose sand, and that it is more than merely "an intellectual dream."

3. In the third place, however, it was possible, and this could not but appear, prima facie, the most probable case, that the attempt might

be frustrated by the difficulties standing in its way, without settling the question, either way, in a perfectly satisfactory manner. But if it were only possible in this way to arrive for oneself at a moderately certain independent judgment upon a matter affecting the highest questions so deeply, even this alone could not but be esteemed a great gain.

Having determined to make the attempt, I had in the first place to decide upon some particular class. The choice was necessarily limited to those the chief forms of which were easily to be obtained alive in some abundance. The Crabs and Macrurous Crustacea, the Stomapoda, the Diastylidae, the Amphipoda and Isopoda, the Ostracoda and Daphnidae, the Copepoda and Parasita, the Cirripedes and Rhizocephala of our coast, representing the class of Crustacea with the deficiency only of the Phyllopoda and Xiphosura, furnished a long and varied, and at the same time intimately connected series, such as was at my command in no other class. But even independently of this circumstance the selection of the Crustacea could hardly have been doubtful. Nowhere else, as has already been indicated by various writers, is the temptation stronger to give to the expressions "relationship, production from a common fundamental form," and the like, more than a mere figurative signification, than in the case of the lower Crustacea. Among the parasitic Crustacea, especially, everybody has long been accustomed to speak, in a manner scarcely admitting of a figurative meaning, of their arrest of development by parasitism, as if the transformation of species were a matter of course. It would certainly never appear to any one to be a pastime worthy of the Deity, to amuse himself with the contrivance of these marvellous cripplings, and so they were supposed to have fallen by their own fault, like Adam, from their previous state of perfection.

That a great part of the larger and smaller groups into which this class is divided, might be regarded as satisfactorily established, was a further advantage not to be undervalued; whilst in two other classes with which I was familiar, namely, the Annelida and Acalephae, all the attempted arrangements could only be considered preliminary revisions. These undisplaceable groups, like the sharply marked forms of the hard, many-jointed dermal framework, were not only important as safe starting points and supports, but were also of the highest value as inflexible barriers in a problem in which, from its very nature, fancy must freely unfold her wings.

When I thus began to study our Crustacea more closely from this new stand-point of the Darwinian theory,—when I attempted to bring their arrangements into the form of a geological tree, and to form some idea of the probable structure of their ancestors,—I speedily saw (as indeed I expected) that it would require years of preliminary work before the essential problem could be seriously handled. The extant systematic works generally laid more weight upon the characters separating the genera, families and orders, than upon those which unite the members of each group, and consequently often furnished but little employable material. But above all things a thorough knowledge of development was indispensable, and every one knows how imperfect is our present knowledge of this subject. The existing deficiencies were the more difficult to supply, because, as Van Beneden remarks with regard to the Decapoda, from the often incredible difference in the development of the most nearly allied forms, these must be separately studied—usually family by family, and frequently genus by genus—nay, sometimes, as in the case of Peneus, even species by species; and because these investigations, in themselves troublesome and tedious, often depend for their success upon a lucky chance.

But although the satisfactory completion of the "Genealogical tree of the Crustacea" appeared to be an undertaking for which the strength and life of an individual would hardly suffice, even under more favourable circumstances than could be presented by a distant island, far removed from the great market of scientific life, far from libraries and museums—nevertheless its practicability became daily less doubtful in my eyes, and fresh observations daily made me more favourably inclined towards the Darwinian theory.

In determining to state the arguments which I derived from the consideration of our Crustacea in favour of Darwin's views, and which (together with more general considerations and observations in other departments), essentially aided in making the correctness of those views seem more and more palpable to me, I am chiefly influenced by an expression of Darwin's: "Whoever," says he ('Origin of Species' page 482), "is led to believe that species are mutable, will do a good service by conscientiously expressing his conviction." To the desire expressed in these words I respond, for my own part, with the more pleasure, as this furnishes me with an opportunity of publicly giving expression in words to the thanks which I feel most deeply to be due from me to Darwin for the instructions and suggestions for which I am so deeply indebted to his book. Accordingly I throw this sand-grain with confidence into the scale against "the load of prejudice by which this subject is overwhelmed," without troubling myself as to whether the priests of orthodox science will reckon me amongst dreamers and children in knowledge of the laws of nature.

CHAPTER 2. THE SPECIES OF MELITA.

A false supposition, when the consequences proceeding from it are followed further and further, will sooner or later lead to absurdities and palpable contradictions. During the period of tormenting doubt—and this was by no means a short one—when the pointer of the scales oscillated before me in perfect uncertainty between the pro and the con, and when any fact leading to a quick decision would have been most welcome to me, I took no small pains to detect some such contradictions among the inferences as to the class of Crustacea furnished by the Darwinian theory. But I found none, either then, or subsequently. Those which I thought I had found were dispelled on closer consideration, or actually became converted into supports for Darwin's theory.

Nor, so far as I am aware, have any of the NECESSARY consequences of Darwin's hypotheses been proved by any one else, to stand in clear and irreconcilable contradiction. And yet, as the most profound students of the animal kingdom are amongst Darwin's opponents, it would seem that it ought to have been an easy matter for them to crush him long since beneath a mass of absurd and contradictory inferences, if any such were to be drawn from his theory. To this want of demonstrated contradictions I think we may ascribe just the same importance in Darwin's favour, that his opponents have attributed to the absence of demonstrated intermediate forms between the species of the various strata of the earth. Independently of the reasons which Darwin gives for the preservation of such intermediate forms being only exceptional, this last mentioned circumstance will not be regarded as of very great significance by any one who has traced the development of an animal upon larvae fished from the sea, and had to seek in vain for months, and even years, for those transitional forms, which he nevertheless knew to be swarming around him in thousands.

A few examples may show how contradictions might come forth as necessary results of the Darwinian hypotheses.

It seems to be a necessity for all crabs which remain for a long time out of the water (but why is of no consequence to us here), that air shall penetrate from behind into the branchial cavity. Now these crabs, which have become more or less estranged from the water, belong to the most different families—the Raninidae (Ranina), Eriphinae (Eriphia gonagra), Grapsoidae (Aratus, Sesarma, etc.), Ocypodidae (Gelasimus, Ocypoda), etc., and the separation of these families must doubtless be referred to a much earlier period than the habit of leaving the water displayed by some of their members. The arrangements connected with aerial respiration, therefore, could not be inherited from a common ancestor, and could scarcely be accordant in their construction. If there were any such accordance not referable to accidental resemblance among them, it would have to be laid in the scale as evidence against the correctness of Darwin's views. I shall show hereafter how in this case the result, far from presenting such contradictions, was rather in the most complete harmony with what might be predicted from Darwin's theory.

(FIGURE 1. Melita exilii n. sp., male, enlarged five times. The large branchial lamellae are seen projecting between the legs.)

A second example.—We are already acquainted with four species of Melita (M. valida, setipes, anisochir, and Fresnelii), and I can add a fifth (Figure 1), in which the second pair of feet bears upon one side a small hand of the usual structure, and on the other an enormous clasp-forceps. This want of symmetry is something so unusual among the Amphipoda, and the structure of the clasp-forceps differs so much from what is seen elsewhere in this order, and agrees so closely in the five species, that one must unhesitatingly regard them as having sprung from common ancestors belonging to them alone among known species. But one of these species, M. Fresnelii, discovered by Savigny, in Egypt, is said to want the secondary flagellum of the anterior antennae, which occurs in the others. From the trustworthiness of all Savigny's works there can scarcely be a doubt as to the correctness of this statement. Now, if the presence or absence of the secondary flagellum possessed the significance of a distinctive generic character, which is usually ascribed to it, or if there were other important differences between Melita Fresnelii and the other species above-mentioned, which would make it seem natural to separate M. Fresnelii as a distinct genus, and to leave the others united with the rest of the species of Melita—that is to say, in the sense of the Darwinian theory, if we assume that all the other Melitae possessed common ancestors, which were not at the same time the ancestors of M. Fresnelii—this would stand in contradiction to the conclusion, derived from the structure of the clasp-forceps, that M. Fresnelii and the four other species above-mentioned possessed common ancestors, which were not also the ancestors of the remaining species of Melita. It would follow:—

1. From the structure of the clasp-forceps: that M. exilii, etc. and M. Fresnelii would branch off together from a stem which branches off from M. palmata.

2. From the presence or absence of the secondary flagellum: that M. palmata, etc. and M. exilii, etc. would branch off together from a stem which branches off from M. Fresnelii.

As, in the first case, among the Crabs, a typical agreement of arrangements produced independently of each other would have been a very suspicious circumstance for Darwin's theory, so also, in the second, would any difference more profound than that of very nearly allied species. Now it seems to me that the secondary flagellum can by no means furnish a reason for doubting the close relationship of M. Fresnelii to M. exilii, etc., which is indicated by the peculiar structure of the unpaired clasp-forceps. In the first place we must consider the possibility that the secondary flagellum, which is not always easy to detect, may only have been overlooked by Savigny, as indeed Spence Bate supposes to have been the case. If it is really deficient it must be remarked that I have found it in species of the genera Leucothoe, Cyrtophium and Amphilochus, in which genera it was missed by Savigny, Dana and Spence Bate—that a species proved by the form of the Epimera (Coxae Sp. B.) of the caudal feet (uropoda Westw.), etc., to be a true Amphithoe* possesses it (* I accept this and all the other genera of Amphipoda here mentioned, with the limits given to them by Spence Bate ('Catalogue of Amphipodous Crustacea').)—that in many species of Cerapus it is reduced to a scarcely perceptible rudiment—nay, that it is sometimes present in youth and disappears (although perhaps not without leaving some trace) at maturity, as was found by Spence Bate to be the case in Acanthonotus Owenii and Atylus carinatus, and I can affirm with regard to an Atylus of these seas, remarkable for its plumose branchiae—and that from all this, at the present day when the increasing number of known Amphipoda and the splitting of them into numerous genera thereby induced, compels us to descend to very minute distinctive characters, we must nevertheless hesitate before employing the secondary flagellum as a generic character. The case of Melita Fresnelii therefore cannot excite any doubts as to Darwin's theory.

CHAPTER 3. MORPHOLOGY OF CRUSTACEA—NAUPLIUS-LARVAE.

If the absence of contradictions among the inferences deduced from them for a narrow and consequently easily surveyed department must prepossess us in favour of Darwin's views, it must be welcomed as a positive triumph of his theory if far-reaching conclusions founded upon it should SUBSEQUENTLY be confirmed by facts, the existence of which science, in its previous state, by no means allowed us to suspect. From many results of this kind upon which I could report, I select as examples, two, which were of particular importance to me, and relate to discoveries the great significance of which in the morphology and classification of the Crustacea will not be denied even by the opponents of Darwin.

Considerations upon the developmental history of the Crustacea had led me to the conclusion that, if the higher and lower Crustacea were at all derivable from common progenitors, the former also must once have passed through Nauplius-like conditions. Soon afterwards I discovered Naupliiform larvae of Shrimps ('Archiv fur Naturgeschichte' 1860 1 page 8), and I must admit that this discovery gave me the first decided turn in Darwin's favour.

The similar number of segments* occurring in the Crabs and Macrura, Amphipoda and Isopoda, in which the last seven segments are always different from the preceding ones in the appendages with which they are furnished, could only be regarded as an inheritance from the same ancestors.

(* Like Claus I do not regard the eyes of the Crustacea as limbs, and therefore admit no ocular segment; on the other hand I count in the median piece of the tail, to which the character of a segment is often denied. In opposition to its interpretation as a segment of the body, only the want of limbs can be cited; in its favour we have the relation of the intestine, which usually opens in this piece, and sometimes even traverses its whole length, as in Microdeutopus and some other Amphipoda. In Microdeutopus, as Spence Bate has already pointed out, one is even led to regard small processes of this tubular caudal piece as rudimentary members. Bell also ('British Stalk-eyed Crustacea' page 20), states that he observed limbs of the last segment in Palaemon serratus in the form of small moveable points.

The attempt has often been made to divide the body of the higher Crustacea into small sections composed of equal numbers of segments, these sections consisting of 3, 5 or 7 segments. None of these attempts has ever met with general acceptance; my own investigations lead me to a conception which nearly approaches Van Beneden's. I assume four sections of 5 segments each—the primitive body, the fore-body, the hind-body, and the middle-body. The primitive body includes the segments which the naupliiform larva brings with it out of the egg; it is afterwards divided, by the younger sections which become developed in its middle, into the head and tail. To this primitive body belong the two pairs of antennae, the mandibles and the caudal feet ("posterior pair of pleopoda," Sp. B.). Even in the mature animal the fact that these terminal sections belong to one another is sometimes betrayed by the resemblance of their appendages, especially that of the outer branch of the caudal feet, with the outer branch (the so-called scale) of the second pair of antennae. Like the antennae, the caudal feet may also become the bearers of high sensorial apparatus, as is shown by the ear of Mysis.

The sequence of the sections of the body in order of time seems originally to have been, that first the fore-body, then the hind-body, and finally the middle-body was formed. The fore-body appears, in the adult animal, to be entirely or partially amalgamated with the head; its appendages (siagonopoda Westw.) are all or in part serviceable for the reception of food, and generally sharply distinguished from those of the following group. The segments of the middle-body seem always to put forth limbs immediately after their own appearance, whilst the segments of the hind-body often remain destitute of feet through long portions of the larval life or even throughout life (as in many female Diastylidae), a reason, among many others, for not, as is usual, regarding the middle-body of the Crustacea as equivalent to the constantly footless abdomen of Insects. The appendages of the middle-body (pereiopoda) seem never, even in their youngest form, to possess two equal branches, a peculiarity which usually characterises the appendages of the hind-body. This is a circumstance which renders very doubtful the equivalence of the middle-body of the Malacostraca with the section of the body which in the Copepoda bears the swimming feet and in the Cirripedia the cirri.

The comprehension of the feet of the hind-body and tail in a single group (as "fausses pattes abdominales," or as "pleopoda") seems not to be justifiable. When there is a metamorphosis, they are probably always produced at different periods, and they are almost always quite different in structure and function. Even in the Amphipoda, in which the caudal feet usually resemble in appearance the last two pairs of abdominal feet, they are in general distinguished by some sort of peculiarity, and whilst the abdominal feet are reproduced in wearisome uniformity throughout the entire order, the caudal feet are, as is well-known, amongst the most variable parts of the Amphipoda.)

And if at the present day the majority of the Crabs and Macrura, and indeed the Stalk-eyed Crustacea in general, pass through Zoea-like developmental states, and the same mode of transformation was to be ascribed to their ancestors, the same thing must also apply, if not to the immediate ancestors of the Amphipoda and Isopoda, at least to the common progenitors of these and the Stalk-eyed Crustacea. Any such assumption as this was, however, very hazardous, so long as not a single fact properly relating to the Edriophthalma could be adduced in its support, as the structure of this very coherent group seemed to be almost irreconcilable with many peculiarities of the Zoea. Thus, in my eyes, this point long constituted one of the chief difficulties in the application of the Darwinian views to the Crustacea, and I could scarcely venture to hope that I might yet find traces of this passage through the Zoea-form among the Amphipoda or Isopoda, and thus obtain a positive proof of the correctness of this conclusion. At this point Van Beneden's statement that a cheliferous Isopod (Tanais Dulongii), belonging, according to Milne-Edwards, to the same family as the common Asellus aquaticus, possesses a carapace like the Decapoda, directed my attention to these animals, and a careful examination proved that these Isopods have preserved, more truly than any other adult Crustacea, many of the most essential peculiarities of the Zoeae, especially their mode of respiration. Whilst in all other Oniscoida the abdominal feet serve for respiration, these in our cheliferous Isopod are solely motory organs, into which no blood-corpuscle ever enters, and the chief seat of respiration is, as in the Zoeae, in the lateral parts of the carapace, which are abundantly traversed by currents of blood, and beneath which a constant stream of water passes, maintained, as in Zoeae and the adult Decapoda, by an appendage of the second pair of maxillae, which is wanting in all other Edriophthalma.

For both these discoveries, it may be remarked in passing, science is indebted less to a happy chance than immediately to Darwin's theory.

Species of Peneus live in the European seas, as well as here, and their Nauplius-brood has no doubt repeatedly passed unnoticed through the hands of the numerous naturalists who have investigated those seas, as well as through my own,* for it has nothing which could attract particular attention amongst the multifarious and often wonderful Nauplius-forms. (* Mecznikow has recently found Naupliiform shrimp-larvae in the sea near Naples.) When I, fancying from the similarity of its movements that it was a young Peneus-Zoea, had for the first time captured such a larva, and on bringing it under the microscope found a Nauplius differing toto coelo from this Zoea, I might have thrown it aside as being completely foreign to the developmental series which I was tracing, if the idea of early Naupliiform stages of the higher Crustacea, which indeed I did not believe to be still extant, had not at the moment vividly occupied my attention.

And if I had not long been seeking among the Edriophthalma for traces of the supposititious Zoea-state, and seized with avidity upon everything that promised to made this refractory Order serviceable to me, Van Beneden's short statement could hardly have affected me so much in the manner of an electric shock, and impelled me to a renewed study of the Tanaides, especially as I had once before plagued myself with them in the Baltic, without getting any further than my predecessors, and I have not much taste for going twice over the same ground.

CHAPTER 4. SEXUAL PECULIARITIES AND DIMORPHISM.

Our Tanais, which in nearly all the particulars of its structure is an extremely remarkable animal, furnished me with a second fact worthy of notice in connection with the theory of the origin of species by natural selection.

When hand-like or cheliform structures occur in the Crustacea, these are usually more strongly developed in the males than in the females, often becoming enlarged in the former to quite a disproportionate size, as we have already seen to be the case in Melita. A better known example of such gigantic chelae is presented by the males of the Calling Crabs (Gelasimus), which are said in running to carry these claws "elevated, as if beckoning with them"—a statement which, however, is not true of all the species, as a small and particularly large-clawed one, which I have seen running about by thousands in the cassava-fields at the mouth of the Cambriu, always holds them closely pressed against its body.

A second peculiarity of the male Crustacea consists not unfrequently in a more abundant development on the flagellum of the anterior antennae of delicate filaments which Spence Bate calls "auditory cilia," and which I have considered to be olfactory organs, as did Leydig before me, although I was not aware of it. Thus they form long dense tufts in the males of many Diastylidae, as Van Beneden also states with regard to Bodotria, whilst the females only possess them more sparingly. In the Copepoda, Claus called attention to the difference of the sexes in this respect. It seems to me, as I may remark in passing, that this stronger development in the males is greatly in favour of the opinion maintained by Leydig and myself, as in other cases male animals are not unfrequently guided by the scent in their pursuit of the ardent females.

Now, in our Tanais, the young males up to the last change of skin preceding sexual maturity resemble the females, but then they undergo an important metamorphosis. Amongst other things they lose the moveable appendages of the mouth even to those which serve for the maintenance of the respiratory current; their intestine is always found empty, and they appear only to live for love. But what is most remarkable is, that they now appear under two different forms. Some acquire powerful, long-fingered, and very mobile chelae, and, instead of the single olfactory filament of the female, have from 12 to 17 of these organs, which stand two or three together on each joint of the flagellum. The others (Figure 5) retain the short thick form of the chelae of the females; but, on the other hand, their antennae (Figure 6) are equipped with a far greater number of olfactory filaments, which stand in groups of from five to seven together.

In the first place, and before inquiring into its significance, I will say a word upon this fact itself. It was natural to consider whether two different species with very similar females and very different males might not perhaps live together, or whether the males, instead of occurring in two sharply defined forms, might not be only variable within very wide limits. I can admit neither of these suppositions. Our Tanais lives among densely interwoven Confervae, which form a coat of about an inch in thickness upon stones in the neighbourhood of the shore. If a handful of this green felt is put into a large glass with clear sea-water, the walls of the glass are soon seen covered with hundreds, nay with thousands, of these little, plump, whitish Isopods. In this way I have examined thousands of them with the simple lens, and I have also examined many hundreds with the microscope, without finding any differences among the females, or any intermediate forms between the two kinds of males.

To the old school this occurrence of two kinds of males will appear to be merely a matter of curiosity. To those who regard the "plan of creation" as the "free conception of an Almighty intellect, matured in the thoughts of the latter before it is manifested in palpable, external forms," it will appear to be a mere caprice of the Creator, as it is inexplicable either from the point of view of practical adaptation, or from the "typical plan of structure." From the side of Darwin's theory, on the contrary, this fact acquires meaning and significance, and it appears in return to be fitted to throw light upon a question in which Bronn saw "the first and most material objection against the new theory," namely, how it is possible that from the accumulation in various directions of the smallest variations running out of one another, varieties and species are produced, which stand out from the primary form clearly and sharply like the petiolated leaf of a

Dicotyledon, and are not amalgamated with the primary form and with each other like the irregular curled lobes of a foliaceous Lichen.

Let us suppose that the males of our Tanais, hitherto identical in structure, begin to vary, in all directions as Bronn thinks, for aught I care. If the species was adapted to its conditions of existence, if the BEST in this respect had been attained and secured by natural selection, fresh variations affecting the species as a species would be retrogressions, and thus could have no prospect of prevailing. They must rather have disappeared again as they arose, and the lists would remain open to the males under variation, only in respect of their sexual relations. In these they might acquire advantages over their rivals by their being enabled either to seek or to seize the females better. The best smellers would overcome all that were inferior to them in this respect, unless the latter had other advantages, such as more powerful chelae, to oppose to them. The best claspers would overcome all less strongly armed champions, unless these opposed to them some other advantage, such as sharper senses. It will be easily understood how in this manner all the intermediate steps less favoured in the development of the olfactory filaments or of the chelae would disappear from the lists, and two sharply defined forms, the best smellers and the best claspers, would remain as the sole adversaries. At the present day the contest seems to have been decided in favour of the latter, as they occur in greatly preponderating numbers, perhaps a hundred of them to one smeller.

To return to Bronn's objection. When he says that "for the support of the Darwinian theory, and in order to explain why many species do not coalesce by means of intermediate forms, he would gladly discover some external or internal principle which should compel the variations of each species to advance in ONE direction, instead of merely permitting them in all directions," we may, in this as in many other cases, find such a principle in the fact that actually only a few directions stand open in which the variations are at the same time improvements, and in which therefore they can accumulate and become fixed; whilst in all others, being either indifferent or injurious, they will go as lightly as they come.

The occurrence of two kinds of males in the same species may perhaps not be a very rare phenomenon in animals in which the males differ widely from the females in structure. But only in those which can be procured in sufficient abundance, will it be possible to arrive at a conviction that we have not before us either two different species, or animals of different ages. From my own observation, although not very extensive, I can give a second example. It relates to a shore-hopper (Orchestia). The animal (Figure 7) lives in marshy places in the vicinity of the sea, under decaying leaves, in the loose earth which the Marsh Crabs (Gelasimus, Sesarma, Cyclograpsus, etc.) throw up around the entrance to their borrows, and even under dry cow-dung and horse-dung. If this species removes to a greater distance from the shore than the majority of its congeners (although some of them advance very far into the land and even upon mountains of a thousand feet in height, such as O. tahitensis, telluris, and sylvicola), its male differs still more from all known species by the powerful chelae of the second pair of feet. Orchestia gryphus, from the sandy coast of Monchgut, alone presents a somewhat similar structure, but in a far less degree; elsewhere the form of the hand usual in the Amphipoda occurs. Now there is a considerable difference between the males of this species, especially in the structure of these chelae—a different so great that we can scarcely find a parallel to it elsewhere between two species of the genus—and yet, as in Tanais, we do not meet with a long series of structures running into one another, but only two forms united by no intermediate terms (Figures 8 and 9). The males would be unhesitatingly regarded as belonging to two well-marked species if they did not live on the same spot, with undistinguishable females. That the two forms of the chelae of the males occur in this species is so far worthy of notice, because the formation of the chelae, which differs widely from the ordinary structure in the other species, indicates that it has quite recently undergone considerable changes, and therefore such a phenomenon was to be expected in it rather than in other species.

I cannot refrain from taking this opportunity of remarking that (so far as appears from Spence Bate's catalogue), for two different kinds of males (Orchestia telluris and sylvicola) which live together in the forests of New Zealand, only one form of female is known, and hazarding the supposition that we have here a similar case. It does not seem to me to be probable that two nearly allied species of these social Amphipoda should occur mixed together under the same conditions of life.

As the males of several species of Melita are distinguished by the powerful unpaired clasp-forceps, the females of some other species of the same genus are equally distinguished from all other Amphipoda by the circumstance that in them a peculiar apparatus is developed which facilitates their being held by the male. The coxal lamellae of the penultimate pair of feet are produced into hook-like processes, of which the male lays hold with the hands of the first pair of feet. The two species in which I am acquainted with this structure are amongst the most salacious animals of their order, even females which are laden with eggs in all stages of development, not unfrequently have their males upon their backs. The two species are nearly allied to Melita palmata Leach (Gammarus Dugesii, Edw.), which is widely distributed on the European coasts, and has been frequently investigated; unfortunately, however, I can find no information as to whether the females of this or any other European species possess a similar contrivance. In M. exilii all the coxal lamellae are of the ordinary formation. Nevertheless, be this as it will, whether they exist in two or in twenty species, the occurrence of these peculiar hook-like processes is certainly very limited.

Now our two species live sheltered beneath slightly tilted stones in the neighbourhood of the shore: one of them, Melita Messalina, so high that it is but rarely covered by the water; the other, Melita insatiabilis, a little lower; both species live together in numerous swarms. We cannot therefore suppose that the loving couples are threatened with disturbance more frequently than those of other species, nor would it be more difficult for the male, than for those of other species, in case of his losing his female, to find a new one. Nor is it any more easy to see how the contrivance on the body of the female for insuring the act of copulation could be injurious to other species. But so long as it is not demonstrated that our species are particularly in want of this contrivance, or that the latter would rather be injurious than beneficial to other species, its presence only in these few Amphipoda will have to be regarded not as the work of far-seeing wisdom, but as that of a favourable chance made use of by Natural Selection. Under the latter supposition its isolated occurrence is intelligible, whilst we cannot perceive why the Creator blessed just these few species with an apparatus which he found to be quite compatible with the "general plan of structure" of the Amphipoda, and yet denied it to others which live under the same external conditions, and equal them even in their extraordinary salacity. Associated with, or in the immediate vicinity of the two species of Melita, live two species of Allorchestes, the pairs of which are met with almost more numerously than the single animals, and yet their females show no trace of the above-mentioned processes of the coxal lamellae.

These cases, I think, must be brought to bear against the conception supported with so much genius and knowledge by Agassiz, that species are embodied thoughts of the Creator; and, with these, all similar instances in which arrangements which would be equally beneficial to all the species of a group are wanting in the majority and only conferred upon a few special favourites, which do not seem to want them any more than the rest.

CHAPTER 5. RESPIRATION IN LAND CRABS.

Among the numerous facts in the natural history of the Crustacea upon which a new and clear light is thrown by Darwin's theory, besides the two forms of the males in our Tanais and in Orchestia Darwinii, there is one which appears to me of particular importance, namely, the character of the branchial cavity in the air-breathing Crabs, of which, unfortunately, I have been unable to investigate some of the most remarkable (Gecarcinus, Ranina). As this character, namely, the existence of an entrance behind the branchiae, has hitherto been noticed, even as a fact, only in Ranina, I will go into it in some detail. I have already mentioned that, as indeed is required by Darwin's theory, this entrant orifice is produced in different manners in the different families.

In the Frog-crab (Ranina) of the Indian Ocean, which, according to Rumphius, loves to climb up on the roofs of the houses, the ordinary anterior entrant orifice is entirely wanting according to Milne-Edwards, and the entrance of a canal opening into the hindmost parts of the branchial cavity is situated beneath the commencement of the abdomen.

The case is most simple in some of the Grapsoidae, as in Aratus Pisonii, a charming, lively Crab which ascends the mangrove bushes (Rhizophora) and gnaws their leaves. By means of its short but remarkably acute claws, which prick like pins when it runs over the hand, this Crab climbs with the greatest agility upon the thinnest twigs. Once, when I had one of these animals sitting upon my hand, I noticed that it elevated the hinder part of its carapace, and that by this means a wide fissure was opened upon each side above the last pair of feet, through which I could look far into the branchial cavity. I have since been unable to procure this remarkable animal again, but on the other hand, I have frequently repeated the same observation upon another animal of the same family (apparently a true Grapsus), which lives abundantly upon the rocks of our coast. Whilst the hinder part of the carapace rises and the above-mentioned fissure is formed, the anterior part seems to sink, and to narrow or entirely close the anterior entrant orifice. Under water the elevation of the carapace never takes place. The animal therefore opens its branchial cavity in front or behind, according as it has to breathe water or air. How the elevation of the carapace is effected I do not know, but I believe that a membranous sac, which extends from the body cavity far into the branchial cavity beneath the hinder part of the carapace, is inflated by the impulsion of the fluids of the body, and the carapace is thereby raised.

I have also observed the same elevation of the carapace in some species of the allied genera Sesarma and Cyclograpsus, which dig deep holes in marshy ground, and often run about upon the wet mud, or sit, as if keeping watch, before their burrows. One must, however, wait for a long time with these animals, when taken out of the water, before they open their branchial cavity to the air, for they possess a wonderful arrangement, by means of which they can continue to breathe water for some time when out of the water. The orifices for the egress of the water which has served for respiration, are situated in these, as in most Crabs, in the anterior angles of the buccal frame ("cadre buccal," M.-Edw.), whilst the entrant fissures of the branchial cavity extend from its hinder angles above the first pair of feet. Now that portion of the carapace which extends at the sides of the mouth between the two orifices ("regions pterygostomiennes"), appears in our animals to be divided into small square compartments. Milne-Edwards has already pointed this out as a particularly remarkable peculiarity. This appearance is caused partly by small wart-like elevations, and partly and especially by curious geniculated hairs, which to a certain extent constitute a fine net or hair-sieve extended immediately over the surface of the carapace. Thus when a wave of water escapes from the branchial cavity, it immediately becomes diffused in this network of hairs and then again conveyed back to the branchial cavity by vigorous movements of the appendage of the outer maxilliped which works in the entrant fissure. Whilst the water glides in this way over the carapace in the form of a thin film, it will again saturate itself with oxygen, and may then serve afresh for the purposes of respiration. In order to complete this arrangement the outer maxillipeds, as indeed has long been known, bear a projecting ridge furnished with a dense fringe of hairs, which commences in front near their median line and passes backwards and outwards to the hinder angle of the buccal frame. Thus the two ridges of the right and left sides form together a triangle with the apex turned forwards,—a breakwater by which the water flowing from the branchial cavity is kept away from the mouth and reconducted to the branchial cavity. In very moist air the store of water contained in the branchial cavity may hold out for hours, and it is only when this is used up that the animal elevates its carapace in order to allow the air to have access to its branchiae from behind.

In Eriphia gonagra the entrant orifices of the respiratory cavity serving for aerial respiration are situated, not, as in the Grapsoidae, above, but behind the last pair of feet at the sides of the abdomen.

The swift-footed Sand-Crabs (Ocypoda) are exclusively terrestrial animals, and can scarcely live for a single day in water; in a much shorter period a state of complete relaxation occurs and all voluntary movements cease.* (* As this was not observed in the sea, but in glass vessels containing sea-water, it might be supposed that the animals become exhausted and die, not because they are under water but because they have consumed all the oxygen which it contained. I therefore put into the same water from which I had just taken an unconscious Ocypoda, with its legs hanging loosely down, a specimen of Lupea diacantha which had been reduced to the same state by being kept in the air, and this recovered in the water just as the Ocypoda did in the air.) In these a peculiar arrangement on the feet of the third and fourth pairs (Figure 12) has long been known, although its connexion with the branchial cavity has not been suspected. These two pairs of feet are more closely approximated than the rest; the opposed surfaces of their basal joints (therefore the hinder surface on the third, and the anterior surface on the fourth feet) are smooth and polished, and their margins bear a dense border of long, silky, and peculiarly formed hairs (Figure 13). Milne-Edwards who rightly compares these surfaces, as to their appearance, with articular surfaces, thinks that they serve to diminish the friction between the two feet. In considering this interpretation, the question could not but arise why such an arrangement for the diminution of friction should be necessary in these particular Crabs and between these two feet, leaving out of consideration the fact that the remarkable brushes of hair, which on the other hand must increase friction, also remain unexplained. But as I was bending the feet of a large Sand-Crab to and fro in various directions, in order to see in what movements of the animal friction occurred at the place indicated, and whether these might, perhaps, be movements of particular importance to it and such as would frequently recur, I noticed, when I had stretched the feet widely apart, in the hollow between them a round orifice of considerable size, through which air could easily be blown into the branchial cavity, and a fine rod might even be introduced into it. The orifice opens into the branchial cavity behind a conical lobe, which stands above the third foot in place of a branchia which is wanting in Ocypoda. It is bounded laterally by ridges, which rise above the articulation of the foot, and to which the lower margin of the carapace is applied. Exteriorly, also, it is overarched by these ridges with the exception of a narrow fissure. This fissure is overlaid by the carapace, which exactly at this part projects further downwards than elsewhere, and in this way a complete tube is formed. Whilst in Grapsus the water is allowed to reach the branchiae only from the front, I saw it in Ocypoda flow in also through the orifice just described.

In the position of posterior entrant orifice and the accompanying peculiarities of the third and fourth pairs of feet, two other non-aquatic species of the same family, which I have had the opportunity of examining, agree with Ocypoda. One of these, perhaps Gelasimus vocans, which lives in the mangrove swamps, and likes to furnish the mouth of its burrow with a thick, cylindrical chimney of several inches in height, has the brushes on the basal joints of the feet in question composed of ordinary hairs. The other, a smaller Gelasimus, not described in Milne-Edwards' 'Natural History of Crustacea,' which prefers drier places and is not afraid to run about on the burning sand under the vertical rays of the noonday sun in December, but can also endure being in water at least for several weeks, resembles Ocypoda in having these brushes composed of non-setiform, delicate hairs, indeed even more delicate and more regularly constructed than in Ocypoda.* (* This smaller Gelasimus is also remarkable because the chameleon-like change of colour exhibited by many Crabs occurs very strikingly in it. The carapace of a male which I have now before me shone with a dazzling white in its hinder parts five minutes since when I captured it, at present it shows a dull gray tint at the same place.) What may be the significance of these peculiar hairs,—whether they only keep foreign bodies from the branchial cavity,—whether they furnish moisture to the air flowing past them,—or whether, as their aspect, especially in the small Gelasimus, reminds one of the olfactory filaments of the Crabs, they may also perform similar functions,—are questions the due discussion of which would lead us too far from our subject. Nevertheless it may be remarked that in both species, especially in Ocypoda, the olfactory filaments in their ordinary situation are very much reduced, and when they are in the water their flagella never perform the peculiar beating movements which may be observed in other Crabs, and even in the larger Gelasimus; moreover, the organ of smell must probably be sought in these air-breathing Crabs, as in the air-breathing Vertebrata, at the entrance to the respiratory cavity.

So much for the facts with regard to the aerial respiration of the Crabs. It has already been indicated why Darwin's theory requires that when any peculiar arrangements exist for aerial respiration, these will be differently constructed in different families. That experience is in perfect accordance with this requirement is the more in favour of Darwin, because the schoolmen far from being able to foresee or explain such profound differences, must rather regard them as extremely surprising. If, in the nearly allied families of the Ocypodidae and Grapsoidae, the closest agreement prevails in all the essential conditions of their structure; if the same plan of structure is slavishly followed in everything else, in the organs of sense, in the articulation of the limbs, in every trabecula and tuft of hairs in the complicated framework of the stomach, and in all the arrangements subserving aquatic respiration, even to the hairs of the flagella employed in cleaning the branchiae,—why have we suddenly this exception, this complete difference, in connection with aerial respiration?

The schoolmen will scarcely have an answer for this question, except by placing themselves on the theologico-teleological stand-point which has justly fallen into disfavour amongst us, and from which the mode of production of an arrangement is supposed to be explained, if its "adaptation" to the animal can be demonstrated. From this point of view we might certainly say that a widely gaping fissure which had nothing prejudicial in it to Aratus Pisonii among the foliage of the mangrove bushes, was not suitable to the Ocypoda living in sand; that in the latter, in order to prevent the penetration of the sand, the orifice of the branchial cavity must be placed at its lowest part, directed downwards, and concealed between broad surfaces fringed with protective brushes of hair. It is far from the intention of these pages to enter upon a general refutation of this theory of adaptation. Indeed there is scarcely anything essential to be added to the many admirable remarks that have been made upon this subject since the time of Spinoza. But this may be remarked, that I regard it as one of the most important services of the Darwinian theory that it has deprived those considerations of usefulness which are still undeniable in the domain of life, of their mystical supremacy. In the case before us it is sufficient to refer to the Gelasimus of the mangrove swamps, which shares the same conditions of life with various Grapsoidae and yet does not agree with them, but with the arenicolous Ocypoda.

CHAPTER 6. STRUCTURE OF THE HEART IN THE EDRIOPHTHALMA.

Scarcely less striking than the example of the air-breathing Crabs, is the behaviour of the heart in the great section Edriophthalma, which may advantageously be divided, after the example of Dana and Spence Bate, only into two orders, the Amphipoda and the Isopoda.

In the Amphipoda, to which the above-mentioned naturalists correctly refer the Caprellidae and Cyamidae (Latreille's Laemodipoda), the heart has always the same position; it extends in the form of a long tube through the six segments following the head, and has three pairs of fissures, furnished with valves, for the entrance of the blood, situated in the second, third, and fourth of these segments. It was found to be of this structure by La Valette in Niphargus (Gammarus puteanus), and by Claus in Phronima; and I have found it to be the same in a considerable number of species belonging to the most different families.* (* The young animals in the egg, a little before their exclusion, are usually particularly convenient for the observation of the fissures in the heart; they are generally sufficiently transparent, the movements of the heart are less violent than at a later period, and they lie still even without the pressure of a glass cover. Considering the common opinion as to the distribution of the Amphipoda, namely, that they increase in multiplicity towards the poles, and diminish towards the equator, it may seem strange that I speak of a considerable number of species on a subtropical coast. I therefore remark that in a few months and without examining any depths inaccessible from the shore, I obtained 38 different species, of which 34 are new, which, with the previously known species (principally described by Dana) gives 60 Brazilian Amphipoda, whilst Kroyer in his 'Gronlands Amfipoder' was acquainted with only 28 species, including 2 Laemodipoda, from the Arctic Seas, although these had been investigated by a far greater number of Naturalists.)

The sole unimportant exception which I have hitherto met with is presented by the genus Brachyscelus,* (*According to Milne-Edwards' arrangement the females of this genus would belong to the "Hyperines ordinaires" and the previously unknown males to the "Hyperines anormales," the distinguishing character of which, namely the curiously zigzagged inferior antennae, is only a sexual peculiarity of the male animals. In systematising from single dead specimens, as to the sex, age, etc. of which nothing is known, similar errors are unavoidable. Thus, in order to give another example of very recent date, a celebrated Ichthyologist, Bleeker, has lately distinguished two groups of the Cyprinodontes as follows: some, the Cyprinodontini, have a "pinna analis non elongata," and the others, the Aplocheilini, a "pinna analis elongata": according to this the female of a little fish which is very abundant here would belong to the first, and the male to the second group. Such mistakes, as already stated, are unavoidable by the "dry-skin" philosopher, and therefore excusable; but they nevertheless prove in how random a fashion the present systematic zoology frequently goes on, without principles or sure foundations, and how much it is in want of the infallible touchstone for the value of the different characters, which Darwin's theory promises to furnish.) in which the heart possesses only two pairs of fissures, as it extends forward only into the second body-segment, and is destitute of the pair of fissures situated in this segment in other forms.* (* I find, in Milne-Edwards' 'Lecons sur la Physiol. et l'Anat. comp.' 3 page 197, the statement that, according to Frey and Leuckart, the heart of Caprella linearis possesses FIVE pairs of fissures. I have examined perfectly transparent young Caprellae (probably the young of Caprella attenuata, Dana, with which they occurred), but can only find the usual three pairs.)

Considering this uniformity presented by the heart in the entire order of the Amphipoda, it cannot but seem very remarkable, that in the very next order of the Isopoda, we find it to be one of the most changeable organs.

In the cheliferous Isopods (Tanais) the heart resembles that of the Amphipoda in its elongated tubular form, as well as in the number and position of the fissures, but with this difference, that the two fissures of each pair do not lie directly opposite each other.

In all other Isopoda the heart is removed towards the abdomen. In the wonderfully deformed parasitic Isopods of the Porcellanae (Entoniscus porcellanae), the spherical heart of the female is confined to a short space of the elongated first abdominal segment, and seems to possess only a single pair of fissures. In the male of

Entoniscus Cancrorum (n. sp.), the heart is situated in the third abdominal segment. In the Cassidinae, the heart (Figure 14) is likewise short and furnished with two pairs of fissures, situated in the last segment of the thorax and the first segment of the abdomen. Lastly, in a young Anilocra, I find the heart (Figure 15) extending through the whole length of the abdomen and furnished with four (or five?) fissures, which are not placed in pairs but alternately to the right and left in successive segments. In other animals of this order, which I have as yet only cursorily examined, further differences will no doubt occur. But why, in two orders so nearly allied to each other, should we find in the one such a constancy, in the other such a variability, of the same highly important organ? From the schoolmen we need expect no explanation, they will either decline the discussion of the "wherefore" as foreign to their province, as lying beyond the boundaries of Natural History, or seek to put down the importunate question by means of a sounding paraphrase of the facts, abundantly sprinkled with Greek words. As I have unfortunately forgotten my Greek, the second way out of the difficulty is closed to me; but as I luckily reckon myself not amongst the incorporated masters, but, to use Baron von Liebig's expression, amongst the "promenaders on the outskirts of Natural History," this affected hesitation of the schoolmen cannot dissuade me from seeking an answer, which indeed presents itself most naturally from Darwin's point of view.

As not only the Tanaides (which reasons elsewhere stated (vide supra) justify us in regarding as particularly nearly related to the primitive Isopod) and the Amphipoda, but also the Decapod Crustacea, possess a heart with three pairs of fissures essentially in the same position; and as the same position of the heart recurs (vide infra) even in the embryos of the Mantis-Shrimps (Squilla), in which the heart of the adult animal, and even, as I have elsewhere shown, that of the larvae when still far from maturity, extends in the form of a long tube with numerous openings far into the abdomen, we must unhesitatingly regard the heart of the Amphipoda as the primitive form of that organ in the Edriophthalma. As, moreover, in these animals the blood flows from the respiratory organs to the heart without vessels, it is very easy to see how advantageous it must be to them to have these organs as much approximated as possible. We have reason to regard as the primitive mode of respiration, that occurring in Tanais (vide supra). Now, where, as in the majority of the Isopoda, branchiae were developed upon the abdomen, the position and structure of the heart underwent a change, as it approached them more nearly, but without the reproduction of a common plan for these earlier modes of structure, either because this transformation of the heart took place only after the division of the primary form into subordinate groups, or because, at least at the time of this division, the varying heart had not yet become fixed in any new form. Where, on the contrary, respiration remained with the anterior part of the body,—whether in the primitive fashion of Zoea, as in the Tanaides, or by the development of branchiae on the thorax, as in the Amphipoda,—the primitive form of the heart was inherited unchanged, because any variations which might make their appearance were rather injurious than advantageous, and disappeared again immediately.

I close this series of isolated examples with an observation which indeed only half belongs to the province of the Crustacea to which these pages ought to be confined, and which also has no further connexion with the preceding circumstances than that of being an "intelligible and intelligence-bringing fact" only from the point of view of Darwin's theory. To-day as I was opening a specimen of Lepas anatifera in order to compare the animal with the description in Darwin's 'Monograph on the Subclass Cirripedia,' I found in the shell of this Cirripede, a blood-red Annelide, with a short, flat body, about half an inch long and two lines in breadth, with twenty-five body-segments, and without projecting setigerous tubercles or jointed cirri. The small cephalic lobe bore four eyes and five tentacles; each body-segment had on each side at the margin a tuft of simple setae directed obliquely upwards, and at some distance from this, upon the ventral surface, a group of thicker setae with a strongly uncinate bidentate apex. There was above EACH of the lateral tufts of bristles a branchia, simple on a few of the foremost segments, and then strongly arborescent to the end of the body. The animal, a female filled with ova, evidently, from these characters, belongs to the family of the Amphinomidae; the only family the members of which, being excellent swimmers, live in the open sea.

That this animal had not strayed accidentally into the Lepas, but appertained to it as a regular and permanent guest, is evidenced by its considerable size in proportion to the narrow entrance of the test of the Lepas, by the complete absence of the iridescence which usually distinguishes the skin of free Annelides and especially of the Amphinomidae, by the formation and position of the inferior setae, etc. But that a worm belonging to this particular family Amphinomidae living in the high sea, occurs as a guest in the Lepas, which also floats in the sea attached to wood, etc., is at once intelligible from the stand-point of the Darwinian theory, whilst the relationship of this parasite to the free-living worms of the open sea remains perfectly unintelligible under the supposition that it was independently created for dwelling in the Lepas.

But however favourable the examples hitherto referred to may be for Darwin, the objection may be raised against them, and that with perfect justice, that they are only isolated facts, which, when the considerations founded upon them are carried far beyond what is immediately given, may only too easily lead us from the right path, with the deceptive glimmer of an ignis fatuus. The higher the structure to be raised, the wider must be the assuring base of well-sifted facts.

Let us turn then to a wider field, that of the developmental history of the Crustacea, upon which science has already brought together a varied abundance of remarkable facts, which, however, have remained a barren accumulation of unmanageable raw-material, and let us see how, under Darwin's hand, these scattered stones unite to form a well-jointed structure, in which everything, bearing and being borne, finds its significant place. Under Darwin's hand! for I shall have nothing to do except just to place the building stones in the position which his theory indicates for them. "When kings build, the carters have to work."

CHAPTER 7. DEVELOPMENTAL HISTORY OF PODOPHTHALMA.

Let us first glance over the extant facts.

Among the Stalk-eyed Crustacea (Podophthalma) we know only a very few species which quit the egg in the form of their parents, with the full number of well-jointed appendages to the body. This is the case according to Rathke* in the European fresh-water Crayfish, and according to Westwood in a West Indian Land Crab (Gecarcinus). (* Authorities are cited only for facts which I have had no opportunity of confirming.) Both exceptions therefore belong to the small number of Stalk-eyed Crustacea which live in fresh water or on the land, as indeed in many other cases fresh-water and terrestrial animals undergo no transformations, whilst their allies in the sea have a metamorphosis to undergo. I may refer to the Earthworms and Leeches among the Annelida, which chiefly belong to the land and to fresh water,—to the Planariae of the fresh waters and the Tetrastemma of the sparingly saline Baltic among the Turbellaria,—to the Pulmonate Gasteropoda, and to the Branchiferous Gasteropoda of the fresh waters, the young of which (according to Troschel's 'Handb. der Zoologie') have no ciliated buccal lobes, although such organs are possessed by the very similar Periwinkles (Littorina).

All the marine forms of this section appear to be subject to a more or less considerable metamorphosis. This appears to be only inconsiderable in the common Lobster, the young of which, according to Van Beneden, are distinguished from the adult animal, by having their feet furnished, like those of Mysis, with a swimming branch projecting freely outwards. From a figure given by Couch the appendages of the abdomen and tail also appear to be wanting.

Far more profound is the difference of the youngest brood from the sexually mature animal in by far the greater majority of the Podophthalma, which quit the egg in the form of Zoea. This young form occurs, so far as our present observations go, in all the Crabs, with the sole exception of the single species investigated by Westwood. I say SPECIES, and not GENUS, for in the same genus, Gecarcinus, Vaughan Thompson found Zoea-brood,* which is also met with in other terrestrial Crabs (Ocypoda, Gelasimus, etc.). (* Bell

('Brit. Stalk-eyed Crust.' page 45) considers himself justified in "eliminating" Thompson's observation at once, because he could only have examined ovigerous females preserved in alcohol. But any one who had paid so much attention as Thompson to the development of these animals, must have been well able to decide with certainty upon eggs, if not too far from maturity or badly preserved, whether a Zoea would be produced from them. Moreover, the mode of life of the Land-Crabs is in favour of Thompson. "Once in the year," says Troschel's 'Handbuch der Zoologie,' "they migrate in great crowds to the sea in order to deposit their eggs, and afterwards return much exhausted towards their dwelling places, which are reached only by a few." For what purpose would be these destructive migrations in species whose young quit the egg and the mother as terrestrial animals?) All the Anomura seem likewise to commence their lives as Zoeae: witness the Porcellanae, the Tatuira (Hippa emerita) and the Hermit Crabs. Among the Macrura we are acquainted with the same earliest form principally in several Shrimps and Prawns, such as Crangon (Du Cane), Caridina (Joly), Hippolyte, Palaemon, Alpheus, etc. Lastly, it is not improbable, that the youngest brood of the Mantis-Shrimps (Squilla) is also in the same case.

The most important peculiarities which distinguish this Zoea-brood from the adult animal, are as follows:—

The middle-body with its appendages, those five pairs of feet to which these animals owe their name of Decapoda, is either entirely wanting, or scarcely indicated; the abdomen and tail are destitute of appendages, and the latter consists of a single piece. The mandibles, as in the Insecta, have no palpi. The maxillipedes, of which the third pair is often still wanting, are not yet brought into the service of the mouth, but appear in the form of biramose natatory feet. Branchiae are wanting, or where their first rudiments may be detected as small verruciform prominences, these are dense cell-masses, through which the blood does not yet flow, and which therefore have nothing to do with respiration. An interchange of the gases of the water and blood may occur all over the thin-skinned surface of the body; but the lateral parts of the carapace may unhesitatingly be indicated as the chief seat of respiration. They consist, exactly as described by Leydig in the Daphniae, of an outer and inner lamina, the space between which is traversed by numerous transverse partitions dilated at their ends; the spaces between these partitions are penetrated by a more abundant flow of blood than occurs anywhere else in the body of the Zoea. To this may be added that a constant current of fresh water passes beneath the carapace in a direction from behind forwards, maintained as in the adult animal, by a foliaceous or linguiform appendage of the second pair of maxillae (Figure 18). The addition of fine coloured particles to the water allows this current of water to be easily detected even in small Zoeae.

The Zoeae of the Crabs (Figure 17) are usually distinguished by long, spiniform processes of the carapace. One of these projects upwards from the middle of the back, a second downwards from the forehead, and frequently there is a shorter one on each side near the posterior inferior angles of the carapace. All these processes are, however, wanting in Maia according to Couch, and in Eurynome according to Kinahan; and in a third species of the same group of the Oxyrhynchi (belonging or nearly allied to the genus Achaeus) I also find only an inconsiderable dorsal spine, whilst the forehead and sides are unarmed. This is another example warning us to be cautious in deductions from analogy. Nothing seemed more probable than to refer back the beak-like formation of the forehead in the Oxyrhynchi to the frontal process of the Zoea, and now it appears that the young of the Oxyrhynchi are really quite destitute of any such process. The following are more important peculiarities of the Zoeae of the Crabs, although less striking than these processes of the carapace which, in combination with the large eyes, often give them so singular an appearance:—the anterior (inner) antennae are simple, not jointed, and furnished at the extremity with from two to three olfactory filaments; the posterior (outer) antennae frequently run out into a remarkably long spine-like process ("styliform process," Spence Bate), and bear, on the outside, an appendage, which is sometimes very minute ("squamiform process" of Spence Bate), corresponding with the antennal scale of the Prawns,* (* In a memoir on the metamorphoses of the Porcellanae I have erroneously described this appendage as the "flagellum.") and the first rudiment of the future flagellum is often already recognisable. Of natatory feet (afterwards maxillipeds) only two pairs are present; the third (not, as Spence Bate thinks, the first) is entirely wanting, or, like the five following pairs of feet, present only as a minute bud. The tail, of very variable form, always bears THREE pairs of setae at its hinder margin. The Zoeae of the Crabs usually maintain themselves in the water in such a manner that the dorsal spine stands upwards, the abdomen is bent forwards, the inner branch of the natatory feet is directed forwards, and the outer one outwards and upwards.

It is further to be remarked that the Zoeae of the Crabs, as also of the Porcellanae, of the Tatuira and of the Shrimps and Prawns, are enveloped, on escaping from the egg, by a membrane veiling the spinous processes of the carapace, the setae of the feet, and the antennae, and that they cast this in a few hours. In Achaeus I have observed that the tail of this earliest larval skin resembles that of the larvae of Shrimps and Prawns, and the same appears to be the case in Maia

Widely as they seem to differ from them at the first glance, the Zoeae of the Porcellanae (Figure 24) approach those of the true Crabs very closely. The antennae, organs of the mouth, and natatory feet, exhibit the same structure. But the tail bears FIVE pairs of setae, and the dorsal spine is wanting, whilst, on the contrary, the frontal process and the lateral spines are of extraordinary length, and directed straight forward and backward.

The Zoea of the Tatuira (Figure 25) also appears to differ but little from those of the true Crabs, which it likewise resembles in its mode of locomotion. The carapace possesses only a short, broad frontal process; the posterior margin of the tail is edged with numerous short setae.

The Zoea of the Hermit Crabs (Figure 26) possesses the simple inner antennae of the Zoea of the true Crabs; the outer antennae bear upon the outside on a short stalk a lamella of considerable size analogous to the scale of the antennae of the Prawns; on the inside, a short, spine-like process; and between the two the flagellum, still short, but already furnished with two apical setae. As in the Crabs, there are only two pairs of well-developed natatory feet (maxillipedes), but the third pair is also present in the form of a two-jointed stump of considerable size, although still destitute of setae.

The tail bears five pairs of setae. The little animal usually holds itself extended straight in the water, with the head directed downwards.

This is also the position in which we usually see the Zoeae of the Shrimps and Prawns (Figure 27), which agree in their general appearance with those of the Hermit Crabs. Between the large compound eyes there is in them a small median eye. The inner antennae bear, at the end of a basal joint sometimes of considerable length, on the inside a plumose seta, which also occurs in the Hermit Crabs, and on the outside a short terminal joint with one or more olfactory filaments. The outer antennae exhibit a well-developed and sometimes distinctly articulated scale, and within this usually a spiniform process; the flagellum appears generally to be still wanting. The third pair of maxillipedes seems to be always present, at least in the form of considerable rudiments. The spatuliform caudal lamina bears from five to six pairs of setae on its hinder margin.

The development of the Zoea-brood to the sexually mature animal was traced by Spence Bate in Carcinus maenas. He proved that the metamorphosis is a perfectly gradual one, and that no sharply separated stages of development, like the caterpillar and pupa of the Lepidoptera, could be defined in it. Unfortunately we possess only this single complete series of observations, and its results cannot be regarded at once as universally applicable; thus the young Hermit Crabs retain the general aspect and mode of locomotion of Zoeae, whilst the rudiments of the thoracic and abdominal feet are growing, and then, when these come into action, appear at once in a perfectly new form, which differs from that of the adult animal chiefly by the complete symmetry of the body and by the presence of four pairs of well-developed natatory feet on the abdomen.* (* Glaucothoe Peronii, M.-Edw., may be a young and still symmetrical Pagurus of this kind.)

The development of the Palinuridiae seems to be very peculiar. Claus found in the ova of the Spiny Lobster (Palinurus), embryos with a completely segmented body, but wanting the appendages of the tail, abdomen, and last two segments of the middle-body; they possess a single median and considerably compound eye; the anterior antennae are simple, the posterior furnished with a small secondary branch; the mandibles have no palpi; the maxillipedes of the third pair, like the two following pairs of feet, are divided into two branches of nearly equal length; whilst the last of the existing pairs of feet and the second pair of maxillipedes bear only an inconsiderable secondary branch. Coste, as is well known, asserts that he has bred young Phyllosomata from the ova of this lobster—a statement that requires further proof, especially as the more recent investigations of Claus upon Phyllosoma by no means appear to be in its favour.

The large compound eyes, which usually soon become moveable, and sometimes stand upon long stalks even in the earliest period, as well as the carapace, which covers the entire fore-body, indicate at once that the position of the larvae hitherto considered, notwithstanding all their differences, is under the Podophthalma. But not a single characteristic of this section is retained by the brood of some Prawns belonging to the genus Peneus or in its vicinity. These quit the egg with an unsegmented ovate body, a median frontal eye, and three pairs of natatory feet, of which the anterior are simple, and the other two biramose—in fact, in the larval form, so common among the lower Crustacea, to which O.F. Muller gave the name of Nauplius. No trace of a carapace! no trace of the paired eyes! no trace of masticating organs near the mouth which is overarched by a helmet-like hood!

In the case of one of these species the intermediate forms which lead from the Nauplius to the Prawn, have been discovered in a nearly continuous series.

The youngest Nauplius (Figure 28) is immediately followed by forms in which a fold of skin runs across the back behind the third pair of feet, and four pairs of stout processes (rudiments of new limbs) sprout forth on the ventral surface. Within the third pair of feet, powerful mandibles are developed.

In a subsequent moult the new limbs (maxillae, and anterior and intermediate maxillipedes) come into action, and in this way the Nauplius becomes a Zoea (Figure 29), agreeing perfectly with the Zoea of the Crabs in the number of the appendages of the body, although very different in form and mode of locomotion and even in many particulars of internal structure. The chief organs of motion are still the two anterior pairs of feet, which are slender and furnished with long setae; the third pair of feet loses its branches, and becomes converted into mandibles destitute of palpi. The labrum acquires a spine directed forward and of considerable size, which occurs in all the Zoeae of allied species. The biramose maxillipedes appear to assist but slightly in locomotion. The forked tail reminds us rather of the forms occurring in the lower Crustacea, especially the Copepoda, than of the spatuliform caudal plate which characterises the Zoeae of Alpheus, Palaemon, Hippolyte, and other Prawns, of the Hermit Crabs, the Tatuira and the Porcellanae. The heart possesses only one pair of fissures, and has no muscles traversing its interior like trabeculae, whilst in other Zoeae two pairs of fissures and an interior apparatus of trabeculae are always distinctly recognisable.

During this Zoeal period the paired eyes, the segments of the middle-body and abdomen, the posterior maxillipedes, the lateral caudal appendages and the stump-like rudiments of the feet of the middle-body are formed (Figure 30). The caudal appendages sprout forth like other limbs freely on the ventral surface, whilst in other Prawns, the Porcellanae, etc., they are produced in the interior of the spatuliform caudal plate.

As the feet of the middle-body come into action, simultaneously with other profound changes, the Zoea passes into the Mysis- or Schizopod-form (Figure 31). The antennae cease to serve for locomotion, their place is taken by the thoracic feet, furnished with long setae, and by the long abdomen which just before was laboriously dragged along as a useless burden, but now, with its powerful muscles, jerks the animal through the water in a series of lively jumps. The anterior antennae have lost their long setae, and by the side of the last (fourth) joint, endowed with olfactory filaments, there appears a second branch, which is at first of a single joint. The previously multi-articulate outer branch of the posterior antennae has become a simple lamella, the antennal scale of the Prawn; beside this appears the stump-like rudiment of the flagellum, probably as a new formation, the inner branch disappearing entirely. The five new pairs of feet are biramose, the inner branch short and simple, the outer one longer, annulated at the end, furnished with long setae, and kept, as in Mysis, in constant whirling motion. The heart acquires new fissures, and interior muscular trabeculae.

During the Mysis-period, the auditory organs in the basal joint of the anterior antennae are formed; the inner branches of the first three pairs of feet are developed into chelae and the two hinder pairs into ambulatory feet; palpi sprout from the mandibles, branchiae on the thorax, and natatory feet on the abdomen. The spine on the labrum becomes reduced in size. In this way the animal gradually approaches the Prawn-form, in which the median eye has become indistinct, the spine of the labrum, and the outer branches of the cheliferous and ambulatory feet have been lost, the mandibular palpi and the abdominal feet have acquired distinct joints and setae, and the branchiae come into action.

In another Prawn, the various larval states of which may be easily recognised as belonging to the same series by the presence of a dark-yellow, sharply-defined spot surrounding the median eye, the youngest Zoea , probably produced from the Nauplius, agrees in all essential particulars with the species just described; its further development is, however, very different, especially in that neither the feet of the middle, nor those of the hind-body are formed simultaneously, and that a stage of development comparable to Mysis in the number and structure of the limbs does not occur.

Traces of the outer maxillipedes make their appearance betimes. Then feet appear upon four segments of the middle-body, and these are biramose on the three anterior segments, and simple, the inner branch being deficient, on the fourth segment. On the inner branches the chelae are developed; the outer branches are lost before an inner branch has made its appearance on the fourth segment The latter again becomes destitute of appendages, so that in this case at an early period four, and at a later only three, segments of the middle-body bear limbs. The fifth segment is still entirely wanting, whilst all the abdominal segments have also acquired limbs, and this one after the other, from before backwards. The adult animal, as shown by the three pairs of chelae, will certainly be very nearly allied to the preceding species.* (* The oldest observed larvae (see Figure 33) are characterised by the extraordinary length of the flagella of the outer antennae, and in this respect resemble the larva of Sergestes found by Claus near Messina .This unusual length of the antennae leads to the supposition that they belong to our commonest Prawn, which is very frequently eaten, and is most nearly allied to Peneus setiferus of Florida. Claus's Acanthosoma (l.c. Figure 13) is like the younger Mysis-form of the larva figured by me in the 'Archiv fur Naturgeschichte,' 1836, Taf 2, Figure 18, and which I am inclined to refer to Sicyonia carinata.)

The youngest larva of the Schizopod genus Euphausia observed by Claus, stands very near the youngest Zoea of our Prawns; but whilst its anterior antennae are already biramose, and it therefore appears to be more advanced, it still wants the middle maxillipedes. In it also Claus found the heart furnished with only a single pair of fissures. Do not Nauplius-like states in this case also precede the Zoea?

The developmental history of Mysis, the near relationship of which with the Shrimps and Prawns has recently again been generally recognised, has been described in detail by Van Beneden. So far as I have tested them I can only confirm his statements. The development of the embryo commences with the formation of the tail! This makes its appearance as a simple lobe, the dorsal surface of which is turned towards and closely applied to that of the embryo. (The young of other Stalk-eyed Crustacea are, as is well known, bent in the egg in such a manner that the ventral surfaces of the anterior and posterior halves of the body are turned towards each other,—in these, therefore, the dorsal, and in Mysis the ventral surface appears convex.) The tail soon acquires the furcate form with which we made acquaintance in the last Prawn-Zoea described. Then two pairs of thick ensiform appendages make their appearance at the opposite end of the body, and behind these a pair of tubercles which are easily overlooked. These are the antennae and mandibles. The egg-membrane now bursts, before any internal organ, or even any tissue, except the cells of the cutaneous layer, is formed. The young animal might be called a Nauplius; but essentially there is nothing but a rough copy of a Nauplius-skin, almost like a new egg-membrane, within which the Mysis is developed. The ten pairs of appendages of the fore- (maxillae, maxillipedes) and middle-body make their appearance simultaneously, as do the five pairs of abdominal feet at a later period. Soon after the young Mysis casts the Nauplius-envelope it quits the brood-pouch of the mother.* (* Van Beneden, who regards the eye-peduncles as limbs, cannot however avoid remarking upon Mysis: "Ce pedicule n'apparait aucunement comme les autres appendices, et parait avoir une autre valeur morphologique.")

For some time, owing to an undue importance being ascribed to the want of a particular branchial cavity, Mysis, Leucifer, and Phyllosoma were referred to the Stomapoda, which are now again limited, as originally by Latreille, to the Mantis-shrimps (Squilla), the Glass-shrimps (Erichthus) and their nearest allies. Of the developmental history of these we have hitherto been acquainted with only isolated fragments. The tracing of the development in the egg is rendered difficult by the circumstance, that the Mantis-shrimps do not, like the Decapoda, carry their spawn about with them, but deposit it in the subterranean passages inhabited by them in the form of thin, round, yellow plates. The spawn is consequently exceedingly difficult to procure, and unfortunately it becomes spoilt in a day when it is removed from its natural hatching place, whilst on the contrary the progress of development may be followed for weeks together in the eggs of a single Crab kept in confinement. The eggs of Squilla, like those removed from the body of the Crab, die because they are deprived of the rapid stream of fresh water which the mother drives through her hole for the purpose of her own respiration.

The accompanying representation of the embryo of Squilla shows that it possesses a long, segmented abdomen without appendages, a bilobate tail, six pairs of limbs, and a short heart; the latter only pulsates weakly and slowly. If it acquires more limbs before exclusion, the youngest larva must stand on the same level as the youngest larva of Euphausia observed by Claus.

Of the two larval forms at present known which are with certainty to be ascribed, if not to Squilla, at least to a Stomapod, I pass over the younger one* (* 'Archiv fur Naturgeschichte' 1863 Taf 1.) as its limbs cannot be positively interpreted, and will only mention that in it the last three abdominal segments are still destitute of appendages. The older larva which resembles the mature Squilla especially in the structure of the great raptorial feet and of the preceding pair, still wants the six pairs of feet following the raptorial feet. The corresponding body-segments are already well developed, an unpaired eye is still present, the anterior antennae are already biramose, whilst the flagellum is wanting in the posterior, and the mandibles are destitute of palpi; the four anterior abdominal segments bear biramose natatory feet, without branchiae; the fifth abdominal segment has no appendages, and this is also the case with the tail, which still appears as a simple lamina, fringed on the hinder margin with numerous short teeth. It is evident that the larva stands essentially in the grade of Zoea.

CHAPTER 8. DEVELOPMENTAL HISTORY OF EDRIOPHTHALMA.

Less varied than that of the Stalk-eyed Crustacea is the mode of development of the Isopoda and Amphipoda, which Leach united in the section Edriophthalma, or Crustacea with sessile eyes.

The Rock-Slaters (Ligia) may serve as an example of the development of the Isopoda. In these, as in Mysis, the caudal portion of the embryo is bent not downwards, but upwards; as in Mysis also, a larval membrane is first of all formed, within which the Slater is developed. In Mysis this first larval skin may be compared to a Nauplius; in Ligia it appears like a maggot quite destitute of appendages, but produced into a long simple tail The egg-membrane is retained longer than in Mysis; it bursts only when the limbs of the young Slater are already partially developed in their full number. The dorsal surface of the Slater is united to the larval skin a little behind the head. At this point, when the union has been dissolved a little before the change of skin, there is a foliaceous appendage, which exists only for a short time, and disappears before the young Slater quits the brood-pouch of the mother.

The young animal, when it begins to take care of itself, resembles the old ones in almost all parts, except one important difference; it possesses only six, instead of seven pairs of ambulatory feet; and the last segment of the middle-body is but slightly developed and destitute of appendages. It need hardly be mentioned that the sexual peculiarities are not yet developed, and that in the males the hand-like enlargements of the anterior ambulatory feet and the copulatory appendages are still deficient.

To the question, how far the development of Ligia is repeated in the other Isopoda, I can only give an unsatisfactory answer. The curvature of the embryo upwards instead of downwards was met with by me as well as by Rathke in Idothea, and likewise in Cassidina, Philoscia, Tanais, and the Bopyridae,—indeed, I failed to find it in none of the Isopoda examined for this purpose. In Cassidina also the first larval skin without appendages is easily detected; it is destitute of the long tail, but is strongly bent in the egg, as in Ligia, and consequently cannot be mistaken for an "inner egg-membrane." This, however, might happen in Philoscia, in which the larval skin is closely applied to the egg-membrane (Figure 38), and is only to be explained as the larval skin by a reference to Ligia and Cassidina. The foliaceous appendage on the back has long been known in the young of the common Water Slater (Asellus).* (* Leydig has compared this foliaceous appendage of the Water Slaters with the "green gland" or "shell-gland" of other crustacea, assuming that the green gland has no efferent duct and appealing to the fact that the two organs occur "in the same place." This interpretation is by no means a happy one. In the first place we may easily ascertain in Leucifer, as was also found to be the case by Claus, that the "green gland" really opens at the end of the process described by Milne-Edwards as a "tubercule auditif" and by Spence Bate as an "olfactory denticle." And, secondly, the position is about as different as it can well be. In the one case a paired gland, opening at the base of the posterior antennae, and therefore on the lower surface of the SECOND segment; in the other an unpaired structure rising in the median line of the back BEHIND THE SEVENTH SEGMENT, ("behind the boundary line of the first thoracic segment," Leydig).) That the last pair of feet of the thorax is wanting in the young of the Wood-lice (Porcellionides, M.-Edw.) and Fish-lice (Cymothoadiens, M.-Edw.) has already been noticed by Milne-Edwards. This applies also to the Box-Slaters (Idothea), to the viviparous Globe-Slaters (Sphaeroma) and Shield-Slaters (Cassidina), to the Bopyridae (Bopyrus, Entoniscus, Cryptoniscus, n.g.), and to the Cheliferous Slaters (Tanais), and therefore probably to the great majority of the Isopoda. All the other limbs are usually well developed in the young Isopoda. In Tanais alone, all the abdominal feet are wanting (but not those of the tail); they are developed simultaneously with the last pair of feet of the thorax.

The last pair of feet on the middle-body of the larva, consequently the penultimate pair in the adult animal, is almost always similar in structure to the preceding pair. A remarkable exception is, however, presented in this respect by Cryptoniscus and Entoniscus,—remarkable as a confirmation of Darwin's proposition that "parts developed in an unusual manner are very variable," for in the peculiarly-formed pair of feet there exists the greatest possible difference between the three species hitherto observed. In Cryptoniscus (Figure 39) this last foot is thin and rod-like; in Entoniscus Cancrorum remarkably long and furnished with a strongly thickened hand and a peculiarly constructed chela; in Entoniscus Porcellanae very short, imperfectly jointed, and with a large ovate terminal joint (Figure 40).

Some Isopods undergo a considerable change immediately before the attainment of sexual maturity. This is the case with the males of Tanais which have already been noticed, and, according to Hesse, with the Pranizae, in which both sexes are said to pass into the form known as Anceus. But Spence Bate, a careful observer, states that he has seen females of the form of Praniza laden with eggs far advanced in their development.

In this order we meet for the first time with an extensive retrograde metamorphosis as a consequence of a parasitic mode of life. Even in some Fish-lice (Cymothoa) the young are lively swimmers, and the adults stiff, stupid, heavy fellows, whose short clinging feet are capable of but little movement. In the Bopyridae (Bopyrus, Phryxus, Kepone, etc., which might have been conveniently left in a single genus), which are parasitic on Crabs, Lobsters, etc., taking up their abode chiefly in the branchial cavity, the adult females are usually quite destitute of eyes; the antennae are rudimentary; the broad body is frequently unsymmetrically developed in consequence of the confined space; its segments are more or less amalgamated with each other; the feet are stunted, and the appendages of the abdomen transformed from natatory feet with long setae into foliaceous or tongue-shaped and sometimes ramified branchiae. In the dwarfish males the eyes, antennae, and feet, are usually better preserved than in the females; but on the other hand all the appendages of the abdomen have not unfrequently disappeared, and sometimes every trace of segmentation. In the females of Entoniscus, which are found in the body-cavity of Crabs and Porcellanae, the eyes, antennae, and buccal organs, the segmentation of the vermiform body, and in one species (Figure 41) the whole of the limbs, disappear almost without leaving a trace; and Cryptoniscus planarioides would almost be regarded as a Flatworm rather than an Isopod, if its eggs and young did not betray its Crustacean nature. Among the males of these various Bopyridae, that of Entoniscus Porcellanae occupies the lowest place; it is confined all its life to six pairs of feet, which are reduced to shapeless rounded lumps.

The Amphipoda are distinguishable from the Isopoda at an early period in the egg by the different position of the embryo, the hinder extremity of which is bent downwards. In all the animals of this order which have been examined for it,* (* In the genera Orchestoidea, Orchestia, Allorchestes, Montagua, Batea n.g., Amphilochus, Atylus, Microdeutopus, Leucothoe, Melita, Gammarus (according to Meissner and La Valette), Amphithoe, Cerapus, Cyrtophium, Corophium, Dulichia, Protella and Caprella.) a peculiar structure makes its appearance very early on the anterior part of the back, by which the embryo is attached to the "inner egg-membrane," and which has been called the "micropylar apparatus," but improperly as it seems to me.* (* Little as a name may actually affect the facts, we ought certainly to confine the name "micropyle" to canals of the egg-membrane, which serve for the entrance of the semen. But the outer egg-membrane passes over the "micropylar apparatus" of the Amphipoda without any perforation, according to Meissner's and La Valette's own statements; it appears never to be present before fecundation, attains its greatest development at a subsequent period of the ovular life, and the delicate canals which penetrate it do not even seem to be always present, indeed it seems to belong to the embryo rather than to the egg-membrane. I have never been able to convince myself that the so-called "inner egg-membrane" is really of this nature, and not perhaps the earliest larva skin, not formed until after impregnation, as might be supposed with reference to Ligia, Cassidina and Philoscia.) It will remind us of the union of the young Isopoda with the larval membrane and of the unpaired "adherent organ" on the nape of the Cladocera, which is remarkably developed in Evadne and persists throughout life; but in Daphnia pulex, according to Leydig, although present in the young animals, disappears without leaving a trace in the adults.

The young animal, whilst still in the egg, acquires the full number of its segments and limbs. In cases where segments are amalgamated together, such as the last two segments of the thorax in Dulichia, the last abdominal segments and the tail in Gammarus ambulans and Corophium dentatum, n. sp., and the last abdominal segments and the tail in Brachyscelus,* or where one or more segments are deficient, as in Dulichia and the Caprellae, we find the same fusion and the same deficiencies in young animals taken out of the brood-pouch of their mother. (* According to Spence Bate, in Brachyscelus crusculum the fifth abdominal segment is not amalgamated with the sixth (the tail) but with the fourth, which I should be inclined to doubt, considering the close agreement which this species otherwise shows with the two species that I have investigated.) Even peculiarities in the structure of the limbs, so far as they are common to both sexes, are usually well-marked in the newly hatched young, so that the latter generally differ from their parents only by their stouter form, the smaller number of the antennal joints and olfactory filaments, and also of the setae and teeth with which the body or feet are armed, and perhaps by the comparatively larger size of the secondary flagellum. An exception to this rule is presented by the Hyperinae which usually live upon Acalephae. In these the young and adults often have a remarkably different appearance; but even in these there is no new formation of body-segments and limbs, but only a gradual transformation of these parts.*

(* In the young of Hyperia galba Spence Bate did not find any of the abdominal feet, or the last two pairs of thoracic feet, but this very remarkable statement required confirmation the more because he examined these minute animals only in the dried state. Subsequently I had the wished-for opportunity of tracing the development of a Hyperia which is not uncommon upon Ctenophora, especially Beroe gilva, Eschsch. The youngest larva from the brood-pouch of the mother already possess THE WHOLE of the thoracic feet; on the other hand, like Spence Bate, I cannot find those of the abdomen. At first simple enough, all these feet soon become converted, like the anterior feet, into richly denticulated prehensile feet, and indeed of three different forms, the anterior feet (Figure 44) the two following pairs (Figure 45) and finally the three last pairs (Figure 46) being similarly constructed and different from the rest. In this form the feet remain for a very long time, whilst the abdominal appendages grow into powerful natatory organs, and the eyes, which at first seemed to me to be wanting, into large hemispheres. In the transition to the form of the adult animal the last three pairs of feet especially undergo a considerable change. The difference between the two sexes is considerable; the females are distinguished by a very broad thorax, and the males (Lestrigonus) by very long antennae, of which the anterior bear an unusual abundance of olfactory filaments.

Their youngest larvae of course cannot swim; they are helpless little animals which firmly cling especially to the swimming laminae of their host; the adult Hyperiae, which are not unfrequently met with free in the sea, are, as is well known, the most admirable swimmers in their order. ("Il nage avec une rapidite extreme," says Van Beneden of H. Latreillii M.-Edw.)

The transformation of the Hyperiae is evidently to be regarded as ACQUIRED and not INHERITED, that is to say the late appearance of the abdominal appendages and the peculiar structure of the feet in the young are not to be brought into unison with the historical development of the Amphipoda, but to be placed to the account of the parasitic mode of life of the young.

As in Brachyscelus, free locomotion has been continued to the adult and not to the young, contrary to the usual method among parasites. Still more remarkable is a similar circumstance in Caligus, among the parasitic Copepoda. The young animal, described by Burmeister as a peculiar genus, Chalimus, lies at anchor upon a fish by means of a cable springing from its forehead, and having its extremity firmly seated in the skin of the fish. When sexual maturity is attained, the cable is cut, and the adult Caligi, which are admirable swimmers, are not unfrequently captured swimming freely in the sea. Thus, in order to give a few examples, the powerful chelae of the antepenultimate pair of feet, of Phromina sedentaria, are produced, according to Pagenstecher, from simple feet of ordinary structure; and vice versa, the chelae on the penultimate pair of feet of the young Brachyscelus, become converted into simple feet. In the young of the last-mentioned genus the long head is drawn out into a conical point and bears remarkably small eyes; in course of growth, the latter, as in most of the Hyperinae, attain an enormous size, and almost entirely occupy the head, which then appears spherical, etc.

The difference of the sexes which, in the Gammarinae is usually expressed chiefly in the structure of the anterior feet (gnathopoda, Sp. Bate) and in the Hyperinae in the structure of the antennae, is often so great that males and females have been described as distinct species, and even repeatedly placed in different genera (Orchestia and Talitrus, Cerapus and Dercothoe, Lestrigonus and Hyperia) or even families (Hyperines anormales and Hyperines ordinaires).

Nevertheless it is only developed when the animals are nearly full-grown. Up to this period the young resemble the females in a general way, even in some cases in which these differ more widely than the males from the "Type" of the order. Thus in the male Shore-hoppers (Orchestia) the second pair of the anterior feet is provided with a powerful hand, as in the majority of the Amphipoda, but very differently constructed in the females. The young, nevertheless, resemble the female. Thus also,—and this is an extremely rare case,* (* "I know of no case in which the inferior (antennae) are obsolete, when the superior are developed," Dana. (Darwin, 'Monograph on the Subclass Cirripedia, Lepadidae' page 15.)—the females of Brachyscelus are destitute of the posterior (or inferior) antennae; the male possesses them like other Amphipodae; in the young I, like Spence Bate, can find no trace of them.

It is, however, to be particularly remarked, that the development of the sexual peculiarities does not stand still on the attainment of sexual maturity.

For example, the younger sexually mature males of Orchestia Tucurauna, n. sp., have slender inferior antennae, with the joints of the flagellum not fused together, the clasping margin ("palm," Sp. Bate) of the hand in the second pair of feet is uniformly convex, the last pair of feet is slender and similar to the preceding. Subsequently the antennae become thickened, two, three, or four of the first joints of the flagellum are fused together, the palm of the hand acquires a deep emargination near its inferior angle, and the intermediate joints of the last pair of feet become swelled into a considerable incrassation. No museum-zoologist would hesitate about fabricating two distinct species, if the oldest and youngest sexually mature males were sent to him without the uniting intermediate forms. In the younger males of Orchestia Tucuratinga, although the microscopic examination of their testes showed that they were already sexually mature, the emargination of the clasping margin of the hand (represented in Figure 50) and the corresponding process of the finger, are still entirely wanting. The same may be observed in Cerapus and Caprella, and probably in all cases where hereditary sexual differences occur.

Next to the extensive sections of the Stalk-eyed and Sessile-eyed Crustacea, but more nearly allied to the former than to the latter, comes the remarkable family of the Diastylidae or Cumacea. The young, which Kroyer took out of the brood-pouch of the female, and which attained one-fourth of the length of their mother, resembled the adult animals almost in all parts. Whether, as in Mysis and Ligia, a transformation occurs within the brood-pouch, which is constructed in the same way as in Mysis, is not known.* (* A trustworthy English Naturalist, Goodsir, described the brood-pouch and eggs of Cuma as early as 1843. Kroyer, whose painstaking care and conscientiousness is recognised with wonder by every one who has met him on a common field of work, confirmed Goodsir's statements in 1846, and, as above mentioned, took out of the brood-pouch embryos advanced in development and resembling their parents. By this the question whether the Diastylidae are full-grown animals or larvae, is completely and for ever set at rest, and only the famous names of Agassiz, Dana and Milne-Edwards, who would recently reduce them again to larvae (see Van Beneden, 'Rech. sur la Fauna littor. de Belgique' Crustacees pages 73 and 74), induce me, on the basis of numerous investigations of my own, to declare in Van Beneden's words; "Parmi toutes les formes embryonnaires de podophthalmes ou d'edriophthalmes que nous avons observees sur nos cotes, nous n'en avons pas vu une seule qui eut meme la moindre resemblance avec un Cuma quelconque." The ONLY THING that suits the larvae of Hippolyte, Palaemon and Alpheus, in the family character of the Cumacea as given by Kroyer which occupies three pages (Kroyer, 'Naturh. Tidsskrift, Ny Raekke,' Bd. 2 pages 203 to 206) is: "Duo antennarum paria." And this, as is well known, applies to nearly all Crustacea. How well warranted are we therefore in identifying the latter with the former. However, it is sufficient for any one to glance at the larva of Palaemon (Figure 27) and the Cumacean (Figure 52) in order to be convinced of their extraordinary similarity!) The caudal portion of the embryo in the Diastylidae, as I have recently observed, is curved upwards as in the Isopoda, and the last pair of feet of the thorax is wanting.

Equally scanty is our knowledge of the developmental history of the Ostracoda. We know scarcely anything except that their anterior limbs are developed before the posterior one (Zenker). The development of Cypris has recently been observed by Claus: — "The youngest stages are shell-bearing Nauplius-forms."

CHAPTER 9. DEVELOPMENTAL HISTORY OF ENTOMOSTRACA, CIRRIPEDES, AND RHIZOCEPHALA.

The section of the Branchiopoda includes two groups differing even in their development,—the Phyllopoda and the Cladocera. The latter minute animals, provided with six pairs of foliaceous feet, which chiefly belong to the fresh waters, and are diffused under similar forms over the whole world, quit the egg with their full number of limbs. The Phyllopoda, on the contrary, in which the number of feet varies between 10 and 60 pairs, and some of which certainly live in the saturated lie of salterns and natron-lakes, but of which only one rather divergent genus (Nebalia) is found in the sea,* have to undergo a metamorphosis. (* If the Phyllopoda may be regarded as the nearest allies of the Trilobites, they would furnish, with Lepidosteus and Polypterus, Lepidosiren and Protopterus, a further example of the preservation in fresh waters of forms long since extinguished in the sea. The occurrence of the Artemiae in supersaline water would at the same time show that they do not escape destruction by means of the fresh water, but in consequence of the less amount of competition in it.) Mecznikow has recently observed the development of Nebalia, and concludes from his observations "that Nebalia, during its embryonal life, passes through the Nauplius- and Zoea-stages, which in the Decapoda occur partly (in Peneus) in the free state." "Therefore," says he, "I regard Nebalia as a Phyllopodiform Decapod." The youngest larvae [of the Phyllopoda] are Nauplii, which we have already met with exceptionally in some Prawns, and which we shall now find reproduced almost without exception. The body-segments and feet, which are sometimes so numerous, are formed gradually from before backwards, without the indication of any sharply-discriminated regions of the body either by the time of their appearance or by their form. All the feet are essentially constructed in the same manner and resemble the maxillae of the higher Crustacea.* We might regard the Phyllopoda as Zoeae which have not arrived at the formation of a peculiarly endowed abdomen or thorax, and instead of these have repeatedly reproduced the appendages which first follow the Nauplius-limbs.

Of the Copepoda—some of which, living in a free state, people the fresh waters, and in far more multifarious forms the sea, whilst others, as parasites, infest animals of the most various classes and often become wonderfully deformed—the developmental history, like their entire natural history, was, until lately, in a very unsatisfactory state. It is true, that we long ago knew that the Cyclopes of our fresh waters were excluded in the Nauplius-form, and that we were acquainted with some others of their young states; we had learnt, through Nordmann, that the same earliest form belonged to several parasitic Crustacea, which had previously passed, almost universally, as worms; but the connecting intermediate forms which would have permitted us to refer the regions of the body and the limbs of the larvae to those of the adult animal, were wanting. The comprehensive and careful investigations of Claus have filled up this deficiency in our knowledge, and rendered the section of the Copepoda one of the best known in the whole class. The following statements are derived from the works of this able naturalist. From the abundance of valuable materials which they contain I select only those which are indispensable for the comprehension of the development of the Crustacea in general, because, in what relates to the Copepoda in particular, the facts have already been placed in the proper light by the representation of their most recent investigator, and must appear to any one whose eyes are open, as important evidence in favour of the Darwinian theory All the larvae of the free Copepoda investigated by Claus, have, at the earliest period, three pairs of limbs (the future antennae and mandibles), the anterior with a single, and the two following ones with a double series of joints, or branchiae. The unpaired eye, labrum, and mouth, already occupy their permanent positions. The posterior portion, which is usually short and destitute of limbs, bears two terminal setae, between which the anus is situated. The form in this Nauplius-brood is extremely various,—it is sometimes compressed laterally, sometimes flat,—sometimes elongated, sometimes oval, sometimes round or even broader than long, and so forth. The changes which the first larval stages undergo during the progress of growth, consist essentially in an extension of the body and the sprouting forth of new limbs. "The following stage already displays a fourth pair of extremities, the future maxillae." Then follow at once three new pairs of limbs (the maxillipedes and the two anterior pairs of natatory feet). The larva still continues like a Nauplius, as the three anterior pairs of limbs represent rowing feet; at the next moult it is converted into the youngest Cyclops-like state, when it resembles the adult animal in the structure of the antennae and buccal organs, although the number of limbs and body segments is still much less, for only the rudiments of the third and fourth pairs of natatory feet have made their appearance in the form of cushions fringed with setae, and the body consists of the oval cephalothorax, the second, third, and fourth thoracic segments, and an elongated terminal joint. In the Cyclopidae the posterior antennae have lost their secondary branch, and the mandibles have completely thrown off the previously existing natatory feet, whilst in the other families these appendages persist, more or less altered. "Beyond this stage of free development, many forms of the parasitic Copepoda, such as Lernanthropus and Chondracanthus, do not pass, as they do not acquire the third and fourth pairs of limbs, nor does a separation of the fifth thoracic segment from the abdomen take place; others (Achtheres) even fall to a lower grade by the subsequent loss of the two pairs of natatory feet. But all free Copepoda, and most of the parasitic Crustacea, pass through a longer or shorter series of stages of development, in which the limbs acquire a higher degree of division into joints in continuous sequence, the posterior pairs of feet are developed, and the last thoracic segment and the different abdominal segments are successively separated from the common terminal portion."

There is only one thing more to be indicated in the developmental history of the parasitic Crustacea, namely that some of them, such as Achtheres percarum, certainly quit the egg like the rest in a Nauplius-like form, inasmuch as the plump, oval, astomatous body bears two pairs of simple rowing feet, and behind these, as traces of the third pair, two inflations furnished each with a long seta, but that beneath this Nauplius-skin a very different larva lies ready prepared, which in a few hours bursts its clumsy envelope and then makes its appearance in a form "which agrees in the segmentation of the body and in the development of the extremities with the first Cyclops-stage" (Claus). The entire series of Nauplius-stages which are passed through by the free Copepoda, are in this case completely over-leapt.

A final and very peculiar section of the Crustacea is formed by the two orders of the Cirripedia and Rhizocephala.* (* The most various opinions prevail as to the position of the Cirripedia. Some ascribe to them a very subordinate position among the Copepoda; as Milne-Edwards (1852). In direct opposition to this notion of his father's, Alph. Milne-Edwards places them (as Basinotes) opposite to all the other Crustacea (Eleutheronotes). Darwin regards them as forming a peculiar sub-class equivalent to the Podophthalma, Edriophthalma, etc. This appears to me to be most convenient. I would not combine the Rhizocephala with the Cirripedia, as Liljeborg has done, but place them in opposition as equivalent, like the Amphipoda and Isopoda. The near relationship of the Cirripedia to the Ostracoda is also spoken of, but the similarity of the so-called "Cypris-like larvae," or Cirriped-pupae as Darwin calls them, to Cypris is so purely external, even as regards the shell, that the relationship appears to me to be scarcely greater than that of Peltogaster socialis (Figure 59) with the family of the sausages.)

In these also the brood bursts out in the Nauplius-form, and speedily strips off its earliest larva-skin which is distinguished by no peculiarities worth noticing. Here also we find again the same pyriform shape of the unsegmented body, the same number and structure of the feet, the same position of the median eye (which, however, is wanting in Sacculina purpurea, and according to Darwin in some species of Lepas), and the same position of the "buccal hood," as in the Nauplii of the Prawns and Copepoda. From the latter the Nauplii of the Cirripedia and Rhizocephala are distinguished by the possession of a dorsal shield or carapace, which sometimes (Sacculina purpurea) projects far beyond the body all round; and they are distinguished not only from other Nauplii, but as far as I know from all other Crustacea, by the circumstance that structures which are elsewhere combined with the two anterior limbs (antennae), here occur separated from them.

The anterior antennae of the Copepoda, Cladocera, Phyllopoda (Leydig, Claus), Ostracoda (at least the Cypridinae), Diastylidae, Edriophthalma, and Podophthalma, with few exceptions relating to terrestrial animals or parasites, bear peculiar filaments which I have already repeatedly mentioned as "olfactory filaments." A pair of similar filaments spring, in the larvae of the Cirripedia and Rhizocephala, directly from the brain.

At the base of the inferior antennae in the Decapoda the so-called "green-gland" has its opening; in the Macrura at the end of a conical process. A similar conical process with an efferent duct traversing it is very striking in most of the Amphipoda. In the Ostracoda, Zenker describes a gland situated in the base of the inferior antennae, and opening at the extremity of an extraordinarily long "spine." In the Nauplii of Cyclops and Cyclopsine, Claus finds pale "shell-glands," which commence in the intermediate pair of limbs (the posterior antennae). On the other hand in the Nauplii of the Cirripedia and Rhizocephala the "shell-glands" open at the ends of conical processes, sometimes of most remarkable length, which spring from the angles of the broad frontal margin, and have been interpreted sometimes as antennae (Burmeister, Darwin) and sometimes as mere "horns of the carapace" (Krohn). The connexion of the "shell-glands" with the frontal horns has been recognised unmistakably in the larvae of Lepas, and indeed the resemblance of the frontal horns with the conical processes on the inferior antennae of the Amphipoda, is complete throughout.* (* In connexion with this it may be mentioned that, in the females of Brachyscelus, in which the posterior antennae are deficient, the conical processes with the canal permeating them are nevertheless retained.)

Notwithstanding their agreement in this important peculiarity, the Nauplii of these two orders present material differences in many other particulars. The abdomen of the young Cirripede is produced beneath the anus into a long tail-like appendage which is furcate at the extremity, and over the anus there is a second long, spine-like process; the abdomen in the Rhizocephala terminates in two short points,—in a "moveable caudal fork, as in the Rotatoria," (O. Schmidt). The young Cirripedes have a mouth, stomach, intestine, and anus, and their two posterior pairs of limbs are beset with multifarious teeth, setae, and hooks, which certainly assist in the inception of nourishment. All this is wanting in the young Rhizocephala. The Nauplii of the Cirripedia have to undergo several moults whilst in that form; the Nauplii of the Rhizocephala, being astomatous, cannot of course live long as Nauplii, and in the course of only a few days they become transformed into equally astomatous "pupae," as Darwin calls them.

The carapace folds itself together, so that the little animal acquires the aspect of a bivalve shell, the foremost limbs become transformed into very peculiar adherent feet ("prehensile antennae," Darwin), and the two following pairs are cast off; like the frontal horns. On the abdomen six pairs of powerful biramose natatory feet with long setae have been formed beneath the Nauplius-skin, and behind these are two short, setigerous caudal appendages (Figure 58).

The pupae of the Cirripedia which are likewise astomatous, agree completely in all these parts with those of the Rhizocephala, even to the minutest details of the segmentation and bristling of the natatory feet;* they are especially distinguished from them by the possession of a pair of composite eyes. (* Compare the figure given by Darwin (Balanidae Plate 30 Figure 5) of the first natatory foot of the pupa of Lepas australis, with that of Lernaeodiscus Porcellanae published in the 'Archiv fur Naturgeschichte' (1863 Taf 3 Figure 5). The sole distinction, that in the latter there are only 3 setae at the end of the outer branch, whilst in the Cirripedia there are 4 on the first and 5 on the following natatory feet, may be due to an error on my part.) Sometimes also traces of the frontal horns seem to persist.* (* Darwin describes as "acoustic orifices" small apertures in the shell of the pupae of the Cirripedia, which, frequently surrounded by a border, are situated, in Lepas pectinata, upon short, horn-like processes. I feel scarcely any hesitation in regarding the apertures as those of the "shell-glands," and the horn-like processes as remains of the frontal horns.)

As the Cirripedia and Rhizocephala now in general resemble each other far more than in their Nauplius-state, this is also the case with the individual members of each of the two orders.

The pupae in both orders attach themselves by means of the adherent feet; those of the Cirripedes to rocks, shells, turtles, drift-wood, ships, etc.,—those of the Rhizocephala to the abdomen of Crabs, Porcellanae, and Hermit Crabs. The carapace of the Cirripedes becomes converted, as is well-known, into a peculiar test, on account of which they were formerly placed among the Mollusca, and the natatory feet grow into long cirri, which whirl nourishment towards the mouth, which is now open. The Rhizocephala remain astomatous; they lose all their limbs completely, and appear as sausage-like, sack-shaped or discoidal excrescences of their host, filled with ova (Figures 59 and 60); from the point of attachment closed tubes, ramified like roots, sink into the interior of the host, twisting round its intestine, or becoming diffused among the sac-like tubes of its liver. The only manifestations of life which persist in these non plus ultras in the series of retrogressively metamorphosed Crustacea, are powerful contractions of the roots, and an alternate expansion and contraction of the body, in consequence of which water flows into the brood-cavity and is again expelled, through a wide orifice.* (* The roots of Sacculina purpurea (Figure 60) which is parasitic upon a small Hermit Crab, are made use of by two parasitic Isopods, namely a Bopyrus and the before mentioned Cryptoniscus planarioides These take up their abode beneath the Sacculina and cause it to die away by intercepting the nourishment conveyed by the roots; the roots, however, continue to grow, even without the Sacculina, and frequently attain an extraordinary extension, especially when a Bopyrus obtains its nourishment from them.)

Out of several Cirripedes, which are anomalous both in structure and development, Cryptophialus minutus must be mentioned here; Darwin found it in great quantities together in the shell of Concholepas peruviana on the Chonos Islands. The egg, which is at first elliptical, soon, according to Darwin, becomes broader at the anterior extremity, and acquires three club-shaped horns, one at each anterior angle and one behind; no internal parts can as yet be detected. Subsequently the posterior horn disappears, and the adherent feet may be recognised within the anterior ones. From this "egg-like larva"—(Darwin says of it, "I hardly know what to call it")—the pupa is directly produced. Its carapace is but slightly compressed laterally and hairy, as in Sacculina purpurea; the adherent feet are of considerable size, and the natatory feet are wanting, as, in the adult animal, are the corresponding cirri. As I learn from Mr. Spence Bate, the Nauplius-stage appears to be overleaped and the larvae to leave the egg in the pupa-form, in the case of a Rhizocephalon found by Dr. Powell in the Mauritius.

I will conclude this general view with a few words upon the earliest processes in the development of the Crustacea. Until recently it was regarded as a general rule that, by the partial segmentation of the vitellus a germinal disc was formed, and in this, corresponding to the ventral surface of the embryo, a primitive band. We now know that in the Copepoda (Claus), in the Rhizocephala (Figure 64), and, as I can add, in the Cirripedia) the segmentation is complete, and the embryos are sketched out in their complete form without any preceding primitive band. Probably the latter will always be the case where the young are hatched as true Nauplii (and not merely with a Nauplius-skin, as in Achtheres). The two modes of development may occur in very closely allied animals, as is proved by Achtheres among the Copepoda.* (* I have not mentioned the Pycnogonidae, because I do not regard them as Crustacea; nor the Xiphosura and Trilobites, because, having never investigated them myself, I knew too little about them, and especially because I am unacquainted with the details of the explanations given by Barrande of the development of the latter. According to Mr. Spence Bate "the young of Trilobites are of the Nauplius-form.")

CHAPTER 10. ON THE PRINCIPLES OF CLASSIFICATION.

Perhaps some one else, more fortunate than myself, may be able, even without Darwin, to find the guiding clue through the confusion of developmental forms, now so totally different in the nearest allies, now so surprisingly similar in members of the most distant groups, which we have just cursorily reviewed. Perhaps a sharper eye may be able, with Agassiz, to make out "the plan established from the beginning by the Creator,"* (* "A plan fully matured in the beginning and undeviatingly pursued;" or "In the beginning His plan was formed and from it He has never swerved in any particular" (Agassiz and Gould, 'Principles of Zoology').) who may have written here, as a Portuguese proverb says "straight in crooked lines."* (* "Deos escrive direito em linhas tortas." To read this remarkable writing we need the spectacles of Faith, which seldom suit eyes accustomed to the Microscope.) I cannot but think that we can scarcely speak of a general plan, or typical mode of development of the Crustacea, differentiated according to the separate Sections, Orders, and Families, when, for example, among the Macrura, the River Crayfish leaves the egg in its permanent form; the Lobster with Schizopodal feet; Palaemon, like the Crabs, as a Zoea; and Peneus, like the Cirripedes, as a Nauplius,—and when, still, within this same sub-order Macrura, Palinurus, Mysis and Euphausia again present different young forms,—when new limbs sometimes sprout forth as free rudiments on the ventral surface, and are sometimes formed beneath the skin which passes smoothly over them, and both modes of development are found in different limbs of the same animal and in the same pair of limbs in different animals,—when in the Podophthalma the limbs of the thorax and abdomen make their appearance sometimes simultaneously, or sometimes the former and sometimes the latter first, and when further in each of the two groups the pairs sometimes all appear together, and sometimes one after the other,—when, among the Hyperina, a simple foot becomes a chela in Phronima and a chela a simple foot in Brachyscelus, etc.

And yet, according to the teaching of the school, it is precisely in youth, precisely in the course of development, that the "Type" is mostly openly displayed. But let us hear what the Old School has to tell us as to the significance of developmental history, and its relation to comparative anatomy and systematic zoology.

Let two of its most approved masters speak.

"Whilst comparative anatomy," said Johannes Muller, in 1844, in his lectures upon this science (and the opinions of my memorable teacher were for many years my own), "whilst comparative anatomy shows us the infinitely multifarious formation of the same organ in the Animal Kingdom, it furnishes us at the same time with the means, by the comparison of these various forms, of recognising the truly essential, the type of these organs, and separating therefrom everything unessential. In this, developmental history serves it as a check or test. Thus, as the idea of development is not that of mere increase of size, but that of progress from what is not yet distinguished, but which potentially contains the distinction in itself, to the actually distinct,—it is clear, that the less an organ is developed, so much the more does it approach the type, and that, during its development, it more and more acquires peculiarities. The types discovered by comparative anatomy and developmental history must therefore agree."

Then, after Johannes Muller has combated the idea of a graduated scale of animals, and of the passage through several animal grades during development, he continues:—"What is true in this idea is, that every embryo at first bears only the type of its section, from which the type of the Class, Order, etc., is only afterwards developed."

In 1856, in an elementary work,* (* 'Principles of Zoology' Part 1 Comparative Physiology. By Louis Agassiz and A.A. Gould Revised Edition Boston 1856.) in which it is usual to admit only what are regarded as the assured acquisitions of science, Agassiz expresses himself as follows:—

"The ovarian eggs of all animals are perfectly identical, small cells with a vitellus, germinal vesicle and germinal spot" . "The organs of the body are formed in the sequence of their organic importance; the most essential always appear first. Thus the organs of vegetative life, the intestine, etc., appear later than those of animal life, the nervous system, skeleton, etc., and these in turn are preceded by the more general phenomena belonging to the animal as such" (paragraph 318). "Thus, in Fishes, the first changes consist in the segmentation of the vitellus and the formation of a germ, processes which are common to all classes of animals. Then the dorsal furrow, characteristic of the Vertebrate, appears—the brain, the organs of the senses; at a later period are formed the intestine, the limbs, and the permanent form of the respiratory organs, from which the class is recognised with certainty. It is only after exclusion that the peculiarities of the structure of the teeth and fins indicate the genus and species" (paragraph 319). "Hence the embryos of different animals resemble each other the more, the younger they are" (paragraph 320). "Consequently the high importance of developmental history is indubitable. For, if the formation of the organs takes place in the order corresponding to their importance, this sequence must of itself be a criterion of their comparative value in classification. The peculiarities which appear earlier should be considered of higher value than those which appear subsequently". "A system, in order to be true and natural, must agree with the sequence of the organs in the development of the embryo"

I do not know whether any one at the present day will be inclined to subscribe to this proposition in its whole extent.* (* Agassiz' own views have lately become essentially different, so far as can be made out from Rud. Wagner's notice of his 'Essay on Classification.' Agassiz himself does not attempt any criticism of the above cited older views, which, however, are still widely diffused. With his recent conception I am unfortunately acquainted only from R. Wagner's somewhat confused report, and have therefore thought it better not to attempt any critical remarks upon it.) It is certain, however, that views essentially similar are still to be met with everywhere in discussions on classification, and that even within the last few years, the very sparingly successful attempts to employ developmental history as the foundation of classification have been repeated.

But how do these propositions agree with our observations on the developmental history of the Crustacea? That these observations relate for the most part to their "free metamorphosis" after their quitting the egg, cannot prejudice their application to the propositions enunciated especially with regard to "embryonal development" in the egg; for Agassiz himself points out (paragraph 391) that both kinds of change are of the same nature and of equal importance and that no "radical distinction" is produced by the circumstance that the former take place before and the latter after birth.

"The ovarian eggs of all animals are identical, small cells with vitellus, germinal vesicle and germinal spot." Yes, somewhat as all Insects are identical, small animals with head, thorax, and abdomen; that is to say if, only noticing what is common to them, we leave out of consideration the difference of their development, the presence or absence and the multifarious structure of the vitelline membrane, the varying composition of the vitellus, the different number and formation of the germinal spots, etc. Numerous examples, which might easily be augmented, of such profound differences, are furnished by Leydig's 'Lehrbuch der Histologie.' In the Crustacea the ovarian eggs actually sometimes furnish excellent characters for the discrimination of species of the same genus; thus, for example, in one Porcellana of this country they are blackish-green, in a second deep blood-red, and in a third dark yellow; and within the limits of the same order they present considerable differences in size, which, as Van Beneden and Claus have already pointed out, stands in intimate connexion with the subsequent mode of development.

"The organs of the body are formed in the sequence of their organic importance; the most essential always appear first." This proposition might be characterised a priori as undemonstrable, since it is impossible either in general, or for any particular animal, to establish a sequence of importance amongst equally indispensable parts. Which is the more important, the lung or the heart—the liver or the kidney?—the artery or the vein? Instead of giving the preference, with Agassiz, to the organs of animal life, we might with equal justice give it to those of vegetative life, as the latter are conceivable without the former, but not the former without the latter. We might urge that, according to this proposition, provisional organs as the first produced must exceed the later-formed permanent organs in importance.

But let us stick to the Crustacea. In Polyphemus Leydig finds the first traces of the intestinal tube even during segmentation. In Mysis a provisional tail is first formed, and in Ligia a maggot-like larva-skin. The simple median eye appears earlier, and would therefore be more important than the compound paired eyes; the scale of the antennae in the Prawns would be more important than the flagellum; the maxillipedes of the Decapoda would be more important than the chelae and ambulatory feet, and the anterior six pairs of feet in the Isopoda, than the precisely similarly formed seventh pair; in the Amphipoda the most important of all organs would be the "micropylar apparatus," which disappears without leaving a trace soon after hatching; in Cyclops the setae of the tail would be more important than all the natatory feet; in the Cirripedia the posterior antennae, as to which we do not know what becomes of them, would be more important than the cirri, and so forth. The most unimportant of all organs would be the sexual organs, and the most essential peculiarity would consist in colour, which is to be referred back to the ovarian egg.

"The embryos, or young states of different animals, resemble each other the more, the younger they are," or, as Johannes Muller expresses it, "they approach the more closely to the common type." Different as may be the ideas connected with the word "type," no one will dispute that the typical form of the penultimate pair of feet in the Amphipoda is that of a simple ambulatory foot, and not that of a chela, for the latter occurs in no single adult Amphipod; we know it only in the young of the genus Brachyscelus, which therefore in this respect undoubtedly depart more widely than the adults from the type of their order. This applies also to the young males of the Shore-hoppers (Orchestia) with regard to the second pair of anterior feet (gnathopoda). In like manner no one will hesitate to accept the possession of seven pairs of feet as a "typical" peculiarity of the Edriophthalma, which Agassiz, on this account, names Tetradecapoda; the young Isopoda, which are Dodecapoda, are also in this respect further from the "type" than the adults.

It is certainly a rule, and this Darwin's theory would lead us to expect, that in the progress of development those forms which are at first similar gradually depart further from each other; but here, as in other classes, the exceptions, for which the Old School has no explanation, are numerous. Not unfrequently we might indeed directly reverse the proposition and assert that the difference becomes the greater, the further we go back in the development, and this not only in those cases in which one of two nearly allied species is directly developed, and the other passes through several larval stages, such as the common Crayfish and the Prawns which are produced from Nauplius-brood. The same may be said, for example, of the Isopoda and Amphipoda. In the adult animals the number of limbs is the same; at the first sight of a Cyrtophium or a Dulichia, and even after the careful examination of a Tanais, we may be in doubt whether we have an Isopod or an Amphipod before us; in the newly-hatched young the number of limbs is different, and if we go back to their existence in the egg, the most passing glance to see whether the curvature is upwards or downwards suffices to distinguish even the youngest embryos of the two orders.

In other instances, the courses which lead from a similar starting-point to a similar goal, separate widely in the middle of the development, as in the Prawns with Nauplius-brood already described.

Finally, so that even the last possibility may be exhausted, it sometimes happens that the greatest similarity occurs in the middle of the development. The most striking example of this is furnished by the Cirripedia and Rhizocephala, whether we compare the two orders or the members of each with one another; from a segmentation quite different in its course (see Figures 61 to 64) proceed different forms of Nauplius, these become converted into exceedingly similar pupae, and from the pupae again proceed sexually mature animals, differing from each other toto coelo.

"If the formation of the organs occurs in the order corresponding to their importance, this sequence must of itself be a criterion of their comparative value in classification." THAT IS TO SAY, SUPPOSING THE PHYSIOLOGICAL AND CLASSIFICATIONAL VALUE OF AN ORGAN TO COINCIDE! Just as in Christian countries there is a catechismal morality, which every one has upon his lips, but no one considers himself bound to follow, or expects to see followed by anybody else, so also has Zoology its dogmas, which are as universally acknowledged, as they are disregarded in practice. Such a dogma as this is the supposition tacitly made by Agassiz. Of a hundred who feel themselves compelled to give their systematic confession of faith as the introduction to a Manual or Monographic Memoir, ninety-nine will commence by saying that a natural system cannot be founded upon a single character, but that it has to take into account all characters, and the general structure of the animal, but that we must not simply sum up these characters like equivalent magnitudes, that we must not count but weigh them, and determine the importance to be ascribed to each of them according to its physiological significance. This is probably followed by a little jingle of words in general terms on the comparative importance of animal and vegetative organs, circulation, respiration, and the like. But when we come to the work itself, to the discrimination and arrangement of the species, genera, families, etc., in all probability not one of the ninety-nine will pay the least attention to these fine rules, or undertake the hopeless attempt to carry them out in detail. Agassiz, for example, like Cuvier, and in opposition to the majority of the German and English zoologists, regards the Radiata as one of the great primary divisions of the Animal Kingdom, although no one knows anything about the significance of the radiate arrangement in the life of these animals, and notwithstanding that the radiate Echinodermata are produced from bilateral larvae. The "true Fishes" are divided by him into Ctenoids and Cycloids, according as the posterior margin of their scales is denticulated or smooth, a circumstance the importance of which to the animal must be infinitely small, in comparison to the peculiarities of the dentition, formation of the fins, number of vertebrae, etc.

And, to return to our Class of the Crustacea, has any particular attention been paid in their classification to the distinctions prevailing in the "most essential organs"? For instance, to the nervous system? In the Corycaeidae, Claus found all the ventral ganglia fused together into a single broad mass, and in the Calanidae a long ventral chain of ganglia,—the former, therefore, in this respect resembling the Spider Crabs and the latter the Lobster; but no one would dream on this account of supposing that there was a relationship between the Corycaeidae and the Crabs, or the Calanidae and the Lobsters.—Or to the organs of circulation? We have among the Copepoda, the Cyclopidae and Corycaeidae without a heart, side by side with the Calanidae and Pontellidae with a heart. And in the same way among the Ostracoda, the Cypridinae, which I find possess a heart, place themselves side by side with Cypris and Cythere which have no such organ.—Or to the respiratory apparatus? Milne-Edwards did this when he separated Mysis and Leucifer from the Decapoda, but he himself afterwards saw that this was an error. In one Cypridina I find branchiae of considerable size, which are entirely wanting in another species, but this does not appear to me to be a reason for separating these species even generically.

On the other hand, what do we know of the physiological significance of the number of segments, and all the other matters which we are accustomed to regard as typical peculiarities of the different organs, and to which we usually ascribe the highest systematic value?

"Those peculiarities which first appear, should be more highly estimated than those which appear subsequently. A system, in order to be true and natural, must agree with the sequence of the organs in the development of the embryo." If the earlier manifested peculiarities are to be estimated more highly than those which afterwards make their appearance, then in those cases in which the structure of the adult animal requires one position in the system, and that of the larva another, the latter and not the former must decide the point. As the Lernaeae and Cirripedes, on account of their Nauplius-brood, were separated from their previous connexions and referred to the Crustacea, we shall, for the same reason, have to separate Peneus from the Prawns and unite it with the Copepoda and Cirripedia. But the most zealous embryomaniac would probably shrink from this course.

A "true and natural system" of the Crustacea to be in accordance with the sequence of the phenomena would have to take into account in the first place the various modes of segmentation, then the position of the embryo, next the number of limbs produced within the egg and so forth, and might be represented somewhat as follows:—

CLASSIS CRUSTACEA.

Sub-class I. HOLOSCHISTA.—Segmentation complete. No primitive band. Nauplius-brood.

Ord. 1. Ceratometopa.—Nauplius with frontal horns. (Cirripedia, Rhizocephala.)

Ord. 2. LEIOMETOPA.—Nauplius without frontal horns. (Copepoda, without Achtheus, etc., Phyllopoda, Peneus.)

Sub-class II. HEMISCHISTA.—Segmentation not complete.

A. Nototropa.—Embryo bent upwards.

Ord. 3. Protura.—The tail is first formed. (Mysis.)

Ord. 3. Saccomorpha.—A maggot-like larva-skin is first formed. (Isopoda.)

B. Gasterotropa.—Embryo bent ventrally.

Ord. 5. Zoeogona.—Full number of limbs not produced in the egg. Zoea-brood. (The majority of the Podophthalmata.)

Ord. 6. Ametabola.—Full number of limbs produced in the egg. (Astacus, Gecarcinus, Amphipoda less Hyperia ?)

This sample may suffice. The farther we go into details in this direction, the more brilliantly, as may easily be imagined, does the naturalness of such an arrangement as this force itself upon us.

All things considered, we may apply the judgment which Agassiz pronounced upon Darwin's theory, with far greater justice to the propositions just examined:—"No theory," says he, "however plausible it may be, can be admitted in science, unless it is supported by facts."

CHAPTER 11. ON THE PROGRESS OF EVOLUTION.

From this scarcely unavoidable but unsatisfactory side-glance upon the old school, which looks down with so great an air of superiority upon Darwin's "intellectual dream" and the "giddy enthusiasm" of its friends, I turn to the more congenial task of considering the developmental history of the Crustacea from the point of view of the Darwinian theory.

Darwin himself, in the thirteenth chapter of his book, has already discussed the conclusions derived from his hypotheses in the domain of developmental history. For a more detailed application of them, however, it is necessary in the first place to trace these general conclusions a little further than he has there done.

The changes by which young animals depart from their parents, and the gradual accumulation of which causes the production of new species, genera, and families, may occur at an earlier or later period of life,—in the young state, or at the period of sexual maturity. For the latter is by no means always, as in the Insecta, a period of repose; most other animals even then continue to grow and to undergo changes. (See above, the remarks on the males of the Amphipoda.) Some variations, indeed, from their very nature, can only occur when the young animal has attained the adult stage of development. Thus the Sea Caterpillars (Polynoe) at first possess only a few body-segments, which, during development, gradually increase to a number which is different in different species, but constant in the same species; now before a young animal could exceed the number of segments of its parents, it must of course have attained that number. We may assume a similar supplementary progress wherever the deviation of the descendants consists in an addition of new segments and limbs.

Descendants therefore reach a new goal, either by deviating sooner or later whilst still on the way towards the form of their parents, or by passing along this course without deviation, but then, instead of standing still, advance still farther.

The former mode will have had a predominant action where the posterity of common ancestors constitutes a group of forms standing upon the same level in essential features, as the whole of the Amphipoda, Crabs, or Birds. On the other hand we are led to the assumption of the second mode of progress, when we seek to deduce from a common original form, animals some of which agree with young states of others.

In the former case the developmental history of the descendants can only agree with that of their ancestors up to a certain point at which their courses separate,—as to their structure in the adult state it will teach us nothing. In the second case the entire development of the progenitors is also passed through by the descendants, and, therefore, so far as the production of a species depends upon this second mode of progress, the historical development of the species will be mirrored in its developmental history. In the short period of a few weeks or months, the changing forms of the embryo and larvae will pass before us, a more or less complete and more or less true picture of the transformations through which the species, in the course of untold thousands of years, has struggled up to its present state.

One of the simplest examples is furnished by the development of the Tubicolar Annelids; but from its very simplicity it appears well adapted to open the eyes of many who, perhaps, would rather not see, and it may therefore find a place here. Three years ago I found on the walls of one of my glasses some small worm-tubes (Figure 65), the inhabitants of which bore three pairs of barbate branchial filaments, and had no operculum. According to this we should have been obliged to refer them to the genus Protula. A few days afterwards one of the branchial filaments had become thickened at the extremity into a clavate operculum (Figure 66), when the animals reminded me, by the barbate opercular peduncle, of the genus Filograna, only that the latter possesses two opercula. In three days more, during which a new pair of branchial filaments had sprouted forth, the opercular peduncle had lost its lateral filaments (Figure 67), and the worms had become Serpulae. Here the supposition at once presents itself that the primitive tubicolar worm was a Protula,—that some of its descendants, which had already become developed into perfect Protulae, subsequently improved themselves by the formation of an operculum which might protect their tubes from inimical intruders,—and that subsequent descendants of these latter finally lost the lateral filaments of the opercular peduncle, which they, like their ancestors, had developed.

What say the schools to this case? Whence and for what purpose, if the Serpulae were produced or created as ready-formed species, these lateral filaments of the opercular peduncle? To allow them to sprout forth merely for the sake of an invariable plan of structure, even when they must be immediately retracted again as superfluous, would certainly be an evidence rather of childish trifling or dictatorial pedantry, than of infinite wisdom. But no, I am mistaken; from the beginning of all things the Creator knew, that one day the inquisitive children of men would grope about after analogies and homologies, and that Christian naturalists would busy themselves with thinking out his Creative ideas; at any rate, in order to facilitate the discernment by the former that the opercular peduncle of the Serpulae is homologous with a branchial filament, He allowed it to make a detour in its development, and pass through the form of a barbate branchial filament.

The historical record preserved in developmental history is gradually EFFACED as the development strikes into a constantly straighter course from the egg to the perfect animal, and it is frequently SOPHISTICATED by the struggle for existence which the free-living larvae have to undergo.

Thus as the law of inheritance is by no means strict, as it gives room for individual variations with regard to the form of the parents, this is also the case with the succession in time of the developmental processes. Every father of a family who has taken notice of such matters, is well aware that even in children of the same parents, the teeth, for example, are not cut or changed, either at the same age, or in the same order. Now in general it will be useful to an animal to obtain as early as possible those advantages by which it sustains itself in the struggle for existence. A precocious appearance of peculiarities originally acquired at a later period will generally be advantageous, and their retarded appearance disadvantageous; the former, when it appears accidentally, will be preserved by natural selection. It is the same with every change which gives to the larval stages, rendered multifarious by crossed and oblique characters, a more straightforward direction, simplifies and abridges the process of development, and forces it back to an earlier period of life, and finally into the life of the egg.

As this conversion of a development passing through different young states into a more direct one, is not the consequence of a mysterious inherent impulse, but dependent upon advances accidentally presenting themselves, it may take place in the most nearly allied animals in the most various ways, and require very different periods of time for its completion. There is one thing, however, that must not be overlooked here. The historical development of a species can hardly ever have taken place in a continuously uniform flow; periods of rest will have alternated with periods of rapid progress. But forms, which in periods of rapid progress were severed from others after a short duration, must have impressed themselves less deeply upon the developmental history of their descendants, than those which repeated themselves unchanged, through a long series of successive generations in periods of rest. These more fixed forms, less inclined to variation, will present a more tenacious resistance in the transition to direct development, and will maintain themselves in a more uniform manner and to the last, however different may be the course of this process in other respects.

In general, as already stated, it will be advantageous to the young to commence the struggle for existence in the form of their parents and furnished with all their advantages—in general, but not without exceptions. It is perfectly clear that a brood capable of locomotion is almost indispensable to attached animals, and that the larvae of sluggish Mollusca, or of worms burrowing in the ground, etc., by swarming briskly through the sea perform essential services by dispersing the species over wider spaces. In other cases a metamorphosis is rendered indispensable by the circumstance that a division of labour has been set up between the various periods of life; for example, that the larvae have exclusively taken upon themselves the business of nourishment. A further circumstance to be taken into consideration is the size of the eggs,—a simpler structure may be produced with less material than a more compound one,—the more imperfect the larva, the smaller may the egg be, and the larger is the number of these that the mother can furnish with the same expenditure of material. As a rule, I believe indeed, this advantage of a more numerous brood will not by any means outweigh that of a more perfect brood, but it will do so in those cases in which the chief difficulty of the young animals consists in finding a suitable place for their development, and in which, therefore, it is of importance to disperse the greatest possible number of germs, as in many parasites.

As the conversion of the original development with metamorphosis into direct development is here under discussion, this may be the proper place to say a word as to the already indicated absence of metamorphosis in fresh-water and terrestrial animals the marine allies of which still undergo a transformation. This circumstance seems to be explicable in two ways. Either species without a metamorphosis migrated especially into the fresh waters, or the metamorphosis was more rapidly got rid of in the emigrants than in their fellows remaining in the sea.

Animals without a metamorphosis would naturally transfer themselves more easily to a new residence, as they had only themselves and not at the same time multifarious young forms to adapt to the new conditions. But in the case of animals with a metamorphosis, the mortality among the larvae, always considerable, must have become still greater under new than under accustomed conditions, every step towards the simplification of the process of development must therefore have given them a still greater preponderance over their fellows, and the effacing of the metamorphosis must have gone on more rapidly. What has taken place in each individual case, whether the species has immigrated after it had lost the metamorphosis, or lost the metamorphosis after its immigration, will not always be easy to decide. When there are marine allies without, or with only a slight metamorphosis, like the Lobster as the cousin of the Cray-fish, we may take up the former supposition; when allies with a metamorphosis still live upon the land or in fresh water, as in the case of Gecarcinus, we may adopt the latter.

That besides this gradual extinction of the primitive history, a FALSIFICATION of the record preserved in the developmental history takes place by means of the struggle for existence which the free-living young states have to undergo, requires no further exposition. For it is perfectly evident that the struggle for existence and natural selection combined with this, must act in the same way, in change and development, upon larvae which have to provide for themselves, as upon adult animals. The changes of the larvae, independent of the progress of the adult animal, will become the more considerable, the longer the duration of the life of the larva in comparison to that of the adult animal, the greater the difference in their mode of life, and the more sharply marked the division of labour between the different stages of development. These processes have to a certain extent an action opposed to the gradual extinction of the primitive history; they increase the differences between the individual stages of development, and it will be easily seen how even a straightforward course of development may be again converted by them into a development with metamorphosis. By this means many, and it seems to me valid reasons may be brought up in favour of the opinion that the most ancient Insects approached more nearly to the existing Orthoptera, and perhaps to the wingless Blattidae, than to any other order, and that the "complete metamorphosis" of the Beetles, Lepidoptera, etc., is of later origin. There were, I believe, perfect Insects before larvae and pupae; but, on the contrary, Nauplii and Zoeae far earlier than perfect Prawns. In contradistinction to the INHERITED metamorphosis of the Prawns, we may call that of the Coleoptera, Lepidoptera, etc. an ACQUIRED metamorphosis.*

(* I will here briefly give my reasons for the opinion that the so-called "complete metamorphosis" of Insects, in which these animals quit the egg as grubs or caterpillars, and afterwards become quiescent pupae incapable of feeding, was not inherited from the primitive ancestor of all Insects, but acquired at a later period.

The order Orthoptera, including the Pseudoneuroptera (Ephemera, Libellula, etc.) appears to approach nearest to the primitive form of Insects. In favour of this view we have:—

1. The structure of their buccal organs, especially the formation of the labium, "which retains, either perfectly or approximately, the original form of a second pair of maxillae" (Gerstacker).

2. The segmentation of the abdomen; "like the labium, the abdomen also very generally retains its original segmentation, which is shown

in the development of eleven segments" (Gerstacker). The Orthoptera with eleven segments in the abdomen, agree perfectly in the number of their body-segments with the Prawn-larva represented in Figure 33, or indeed, with the higher Crustacea (Podophthalma and Edriophthalma) in general, in which the historically youngest last thoracic segment (see page 123), which is sometimes late-developed, or destitute of appendages, or even deficient, is still wanting.

3. That, as in the Crustacea, the sexual orifice and anus are placed upon different segments; "whilst the former is situated in the ninth segment, the latter occurs in the eleventh" (Gerstacker).

4. Their palaeontological occurrence; "in a fossil state the Orthoptera make their appearance the earliest of all Insects, namely as early as the Carboniferous formation, in which they exceed all others in number" (Gerstacker).

5. The absence of uniformity of habit at the present day in an order so small when compared with the Coleoptera, Hymenoptera, etc. For this also is usually a phenomenon characteristic of very ancient groups of forms which have already overstepped the climax of their development, and is explicable by extinction in mass. A Beetle or a Butterfly is to be recognised as such at the first glance, but only a thorough investigation can demonstrate the mutual relationships of Termes, Blatta, Mantis, Forficula, Ephemera, Libellula, etc. I may refer to a corresponding remarkable example from the vegetable world: amongst Ferns the genera Aneimia, Schizaea and Lygodium, belonging to the group Schizaeaceae which is very poor in species, differ much more from each other than any two forms of the group Polypodiaceae which numbers its thousands of species.

If, from all this, it seems right to regard the Orthoptera as the order of Insects approaching most nearly to the common primitive form, we must also expect that their mode of development will agree better with that of the primitive form, than, for example, that of the Lepidoptera, in the same way that some of the Prawns (Peneus) approaching most closely the primitive form of the Decapoda, have most truly preserved their original mode of development. Now, the majority of the Orthoptera quit the egg in a form which is distinguished from that of the adult Insect almost solely by the want of wings; these larvae then soon acquire rudiments of wings, which appear more strongly developed after every moult. Even this perfectly gradual transition from the youngest larva to the sexually mature Insect, preserves in a far higher degree the picture of an original mode of development, than does the so-called complete metamorphosis of the Coleoptera, Lepidoptera, or Diptera, with its abruptly separated larva-, pupa- and imago-states.

The most ancient Insects would probably have most resembled these wingless larvae of the existing Orthoptera. The circumstance that there are still numerous wingless species among the Orthoptera, and that some of these (Blattidae) are so like certain Crustacea (Isopods) in habit that both are indicated by the same name ("Baratta") by the people in this country, can scarcely be regarded as of any importance.

The contrary supposition that the oldest Insects possessed a "complete metamorphosis," and that the "incomplete metamorphosis" of the Orthoptera and Hemiptera is only of later origin, is met by serious difficulties. If all the classes of Arthropoda (Crustacea, Insecta, Myriopoda and Arachnida) are indeed all branches of a common stem (and of this there can scarcely be a doubt), it is evident that the water-inhabiting and water-breathing Crustacea must be regarded as the original stem from which the other terrestrial classes, with their tracheal respiration, have branched off. But nowhere among the Crustacea is there a mode of development comparable to the "complete metamorphosis" of the Insecta, nowhere among the young or adult Crustacea are there forms which might resemble the maggots of the Diptera or Hymenoptera, the larvae of the Coleoptera, or the caterpillars of the Lepidoptera, still less any bearing even a distant resemblance to the quiescent pupae of these animals. The pupae, indeed, cannot at all be regarded as members of an original developmental series, the individual stages of which represent permanent ancestral states, for an animal like the mouthless and footless pupa of the Silkworm, enclosed by a thick cocoon, can never have formed the final, sexually mature state of an Arthropod.

In the development of the Insecta we never see new segments added to those already present in the youngest larvae, but we do see segments which were distinct in the larva afterwards become fused together or disappear. Considering the parallelism which prevails throughout organic nature between palaeontological and embryonic development, it is therefore improbable that the oldest Insects should have possessed fewer segments than some of their descendants. But the larva of the Coleoptera, Lepidoptera, etc., never have more than nine abdominal segments, it is therefore not probable that they represent the original young form of the oldest Insects, and that the Orthoptera, with an abdomen of eleven segments, should have been subsequently developed from them.

Taking into consideration on the one hand these difficulties, and on the other the arguments which indicate the Orthoptera as the order most nearly approaching the primitive form, it is my opinion that the "incomplete metamorphosis" of the Orthoptera is the primitive one, INHERITED from the original parents of all Insects, and the "complete metamorphosis" of the Coleoptera, Diptera, etc., a subsequently ACQUIRED one.)

Which of the different modes of development at present occurring in a class of animals may claim to be that approaching most nearly to the original one, is easy to judge from the above statements.

The primitive history of a species will be preserved in its developmental history the more perfectly, the longer the series of young states through which it passes by uniform steps; and the more truly, the less the mode of life of the young departs from that of the adults, and the less the peculiarities of the individual young states can be conceived as transferred back from later ones in previous periods of life, or as independently acquired.

Let us apply this to the Crustacea.

CHAPTER 12. PROGRESS OF EVOLUTION IN CRUSTACEA.

According to all the characters established in the last paragraph, the Prawn that we traced from the Nauplius through states analogous to Zoea and Mysis to the form of a Macrurous Crustacean appears at present to be the animal, which in the section of the higher Crustacea (Malacostraca) furnishes the truest and most complete indications of its primitive history. That it is the most complete is at once evident. That it is the truest must be assumed, in the first place, because the mode of life of the various ages is less different than in the majority of the other Podophthalma; for from the Nauplius to the young Prawn they were found swimming freely in the sea, whilst Crabs, Porcellanae, the Tatuira, Squilla, and many Macrura, when adult usually reside under stones, in the clefts of rocks, holes in the earth, subterranean galleries, sand, etc., not to mention other deviations in habits such as are presented by the Hermit Crabs, Pinnotheres, etc.,—and secondly and especially because the peculiarities which distinguish the Zoea of this species particularly from other Zoeae (the employment of the anterior limbs for swimming, the furcate tail, the simple heart, the deficiency of the paired eyes and abdomen at first, etc.) are neither to be deduced from a retro-transfer of late-acquired advantages to this early period of life, nor to be regarded at all as advantages over other Zoeae which the larva might have acquired in the struggle for existence.

A similar development must have been once passed through by the primitive ancestor of all Malacostraca, probably differing from that of our Prawn, especially in the circumstance that it would go on more uniformly without the sudden change of form and mode of locomotion produced in the latter by the simultaneous sprouting forth and entering into action in the Nauplius of four and in the Zoea of five pairs of limbs. It is to be supposed that, not only originally but even still, in the larvae of the first Malacostraca, the new body-segments and pairs of limbs are formed singly,—first of all the segments of the fore-body, then those of the abdomen, and finally those of the middle-body,—and, moreover, that in each region of the body the anterior segments were formed earlier than the posterior ones, and therefore last of all the hindermost segment of the middle-body. Of this original mode more or less distinct traces still remain, even in species in which, in other respects, the course of development of their ancestors is already nearly effaced. Thus the abdominal feet of the Prawn-larva represented in Figure 33, are formed singly from before backwards, and after these the last feet of the middle-body; thus, in Palinurus, the last two pairs of feet of the middle-body are formed later than the rest; thus in the young larvae of the Stomapoda the last three abdominal segments are destitute of limbs, which are still wanting on the last of them in older larvae; and thus, in the Isopoda, the historically newest pair of feet is produced later than all the rest. In the Copepoda this formation of new segments and limbs, gradually advancing from before backwards, is more perfectly preserved than in any of the higher Crustacea.* (* It is well known that, in many cases, even in adult animals the last segment of the middle-body, or some of its last segments, either want their limbs or are themselves deficient (Entoniscus Porcellanae male, Leucifer, etc.). This might be due to the animals having separated from the common stem before these limbs were formed at all. But in those cases with which I am best acquainted, it seems to me more probable that the limbs have been subsequently lost again. That these particular limbs and segments are more easily lost than others is explained by the circumstance that, as the youngest, they have been less firmly fixed by long-continued inheritance. ("Mr. Dana believes, that in ordinary Crustaceans, the abortion of the segments with their appendages almost always takes place at the posterior end of the cephalothorax."—Darwin, Balanidae, page 111.))

The original development of the Malacostraca starting from the Nauplius, or the lowest free-living grade with which we are acquainted in the class of Crustacea, is now-a-days nearly effaced in the majority of them. That this extinction has actually taken place in the way already deduced as a direct consequence from Darwin's theory, will be the more easily demonstrated, the more this process is still included in the course of life, and the less completely it is already worn out. We may hope to obtain the most striking examples in the still unknown developmental history of the various Schizopoda, Peneidae, and, indeed, of the Macrura in general. At present the multifarious Zoea-forms appear to be particularly instructive. Almost all the peculiarities by which they depart from the primitive form of the Zoea of Peneus (Figures 29, 30 and 32), may in fact be conceived as transferred back from a later period into this early period of life. This is the case with the large compound eyes,—with the structure of the heart,—with the raptorial feet in Squilla,—and with the powerful, muscular, straightly-extended abdomen in Palaemon, Alpheus, Hippolyte, and the Hermit Crabs. (In the latter, indeed, the abdomen of the adult animal is a shapeless sac filled with the liver and generative organs, but it is still tolerably powerful in the Glaucothoe-stage, and was certainly still more powerful when this stage was still the permanent form of the animal.) It is also the case with the abdomen of the Zoeae of the Crabs, the Porcellanae, and the Tatuira, which is still powerful, although usually bent under the breast; the two last swim tolerably by means of the abdomen, even when adult, as do the true Crabs in the young state known as Megalops. It is the case, lastly, with the conversion of the two anterior pairs of limbs into antennae. The second pair of antennae, which, in the various Zoeae always remains a step behind that of the adult animal, is particularly remarkable. In the Crabs the "scale" is entirely wanting; their Zoeae have it indicated in the form of a moveable appendage, which is often exceedingly minute. In the Hermit Crabs a similar, usually moveable, spiniform process occurs as the remains of the scale; their Zoeae have a well-developed but inarticulate scale. A precisely similar scale is possessed by the adult Prawns, in the Zoeae of which it exists still in a jointed form, like the outer branch of the second pair of feet of the Nauplius or Peneus-Zoea.

The long, spiniform processes on the carapace of the Zoeae of the Crabs and Porcellanae are not to be explained in this way, but their advantage to the larvae is evident. Thus, for example, if the body of the Zoea of Porcellana stellicola , without the processes of the carapace and without the abdomen, which however is not rigidly extensible, is scarcely half a line in length, whilst with the processes it is four lines long, a mouth of eight times the width is necessary in order to swallow the little animal when thus armed.* (* Persephone, a rare Crab, belonging to the family Leucosiidae, is served in the same manner by its long chelate feet. If we seize the animal, it extends them most obstinately straight downwards, so that in all probability we should more easily break than bend them.) Consequently these processes of the carapace may be regarded as acquired by the Zoea itself in the struggle for existence.

The formation of new limbs beneath the skin of the larvae is also to be referred to an earlier occurrence of processes which originally took place at a later period. The original course must have been that they sprouted forth in a free form upon the ventral surface of the larva in the next stage after the change of skin; whilst now they are developed before the change of skin, and thus only come into action a stage earlier. In larvae which, for other reasons, must be regarded as more nearly approaching the primitive form, the original mode usually prevails in this particular also. Thus the caudal feet (the "lateral caudal lamellae") are formed freely on the ventral surface in Euphausia and the Prawns with Nauplius-brood, and within the caudal lamellae in the Prawns with Zoea-brood, in Pagurus and Porcellana.

A compression of several stages into one, and thereby an abridgement and simplification of the course of development, is expressed in the simultaneous appearance of several new pairs of limbs.

How earlier young states may gradually be completely lost, is shown by Mysis and the Isopoda. In Mysis there is still a trace of the Nauplius-stage; being transferred back to a period when it had not to provide for itself, the Nauplius has become degraded into a mere skin; in Ligia this larva-skin has lost the last traces of limbs, and in Philoscia it is scarcely demonstrable.

Like the spinous processes of the Zoeae, the chelae on the penultimate pair of feet of the young Brachyscelus are to be regarded as acquired by the larva itself. The adult animals swim admirably and are not confined to their host; as soon as the specimens of Chrysaora Blossevillei, Less., or Rhizostoma cruciatum, Less., on which they are seated, become the sport of the waves in the neighbourhood of the shore, they escape from them, and are only to be obtained from lively Acalephs. The young are helpless creatures and bad swimmers; a special apparatus for adhesion must be of great service to them.

To review the developmental history of the different Malacostraca in detail would furnish no results at all correspondent to the time occupied by it,—if our knowledge was more complete it would be more profitable. I therefore abandon it, but will not omit to mention that in it many difficulties which cannot at present be satisfactorily solved would present themselves. To these isolated difficulties I ascribe the less importance, however, because even a little while ago, before the discovery of the Prawn-Nauplius, this entiredomain of the development of the Malacostraca was almost inaccessible to Darwin's theory.

Nor will I dwell upon the contradictions which appear to result from the application of the Darwinian theory to this department. I leave it to our opponents to find them out. Most of them may easily be proved to be only apparent. There are two of these objections, however, which lie so much on the surface that they can hardly escape being brought forward, and these, I think, I must get rid of.

"The peculiarities in which the Zoeae of the Crabs, the Porcellanae, the Tatuira, the Hermit Crabs, and the Prawns with Zoea-brood agree, and by which they are in common distinguished from the larvae of Peneus produced from Nauplii, forces us (it might be said) to the supposition that the common ancestor of these various Decapods quitted the egg in a similar Zoea-form. But then neither Peneus with its Nauplius-brood, nor even apparently the Palinuri could be referred back to this ancestor. The mode of development of Peneus and Palinurus, as also several peculiar larvae of unknown origin, but which are in all probability to be attributed to Macrurous Crustacea, necessitate on the contrary the opposite supposition, namely, that the different groups of the Macrura have passed from their original to their present mode of development independently of each other and also independently of the Crabs." To this we may answer that the occurrence of the Zoea-form in all the above-mentioned Decapoda, its existence in Peneus during the whole of that period of life which is richest in progress and in which the wide gap between the Nauplius and the Decapod is filled up, its recurrence even in the development of the Stomapoda, the occurrence of a larval form closely approaching the youngest Zoea of Peneus in the Schizopod genus Euphausia, and the reminiscence of the structure of Zoea, which even the adult Tanais has preserved in its mode of respiration,—all indicate Zoea as one of those steps in development which persisted as a permanent form throughout a long period of repose, perhaps through a whole series of geological formations, and thus has also made a deeper impression upon the development of its descendants, and formed a firmer nucleus in the midst of other and more readily effaced young states. It cannot, therefore, surprise us that in transitions from the original mode of metamorphosis to direct development, even when produced independently, the larval life commences in the same way with this Zoea-form in different families, in which the earlier stages of development are effaced. But except what is common to all Zoeae, and what may easily be explained as being transferred back from a later into this stage, the Zoeae of the Crabs, for example, agree with those of Pagurus and Palaemon in no single peculiarity of structure which leads us to suppose a common inheritance. Consequently we may apparently assume, without hesitation, that when the Brachyura and Macrura separated, the primitive ancestors of each of these groups passed through a more complete metamorphosis, and that the transition to the present mode of development belongs to a later period. With regard to the Brachyura, it may be added that in them this transition occurred only a little later and indeed before the existing families separated. The arrangement of the processes of the carapace, and, still more, the similar number of the caudal setae in the most different Zoeae of Crabs prove this. Such an accordance in the number of organs apparently so unimportant is only explicable by common inheritance. We may predict with certainty that amongst the Brachyura no species will occur which, like Peneus, still produces Nauplius-brood.* (* I must not omit remarking that what has been said as to the development of the Crabs applies essentially only to the groups Cyclometopa, Catometopa and Oxyrhyncha, placed together by Alph. Milne-Edwards as "Eustomes." Among the Oxystomata, as also among the "Anomura apterura," Edw., which approach so nearly to the Crabs, I am unacquainted with the earliest young states of any of the species.)

As we have already seen, Mysis and the Isopoda depart from all other Crustacea very remarkably by the fact that their embryos are curved upwards, instead of, as elsewhere, downwards. Does not so isolated a phenomenon as this, it might be asked, in the sense of Darwin's theory, indicate a common inheritance? Does it not necessitate that we should unite as the descendants of the same primitive ancestors, Mysis with the Isopoda on the one hand, and on the other the rest of the Podophthalma with the Amphipoda? I think not. Such a necessity exists only for those who estimate a peculiarity at a higher value because it makes its appearance at an earlier period of the egg-life. Whoever regards species as not created independently and unchangeably, but as having gradually become what they are, will say to himself that, when the ancestors of our Mysides came (probably much later than those of the Amphipoda and Isopoda) to develop numerous body-segments and limbs whilst still embryos, as they could no longer find room in the egg when extended straight out, and were therefore compelled to bend themselves, this could only take place either upwards or downwards, and whatever conditions may have decided the direction actually adopted, any near relationship to either of the two orders of Edriophthalma could hardly have taken part in it.

It may, however, be remarked, that the different curvature of the embryo in the Amphipoda and Isopoda is so far instructive, as it proves that their present mode of development was adopted only after the separation of these orders, and that, in the primitive stock of the Edriophthalma, the embryos were, if not Nauplii, at least short enough in the body to find room in the egg in an extended position, like the larvae of Achtheres enclosed by the Nauplius-skin. On the other hand the uniformity of development that prevails in each of the two orders—which is expressed in the Amphipoda for example in the formation of the "micropylar apparatus," in the Isopoda in the want of the last pair of ambulatory feet—testifies that the present mode of development has come down from a very early period and extends back beyond the separation of the present families. In these two orders also, as well as in the Crabs, we can hardly hope to find traces of earlier young states, unless it be in the family of the Tanaidae.* (* Whether the want of the abdominal feet in the young of Tanais be an inheritance from the time of the primitive Isopoda, or a subsequently acquired peculiarity, which appears to me the more admissible view at present, may perhaps be decided with some certainty, when we become acquainted with the development and mode of life of its family allies, Apseudes and Rhoea. The latter, as is well known, is the only Isopod which possesses a secondary flagellum on the anterior antennae. I have recently obtained a new and unexpected proof that the Tanaidae ("Asellotes heteropodes" M.-Edw.) of all known Crustacea approach most closely to the primitive form of the Edriophthalma. Mr. C. Spence Bate writes to me: "Apseudes, as far as I know, is the ONLY

Isopod in which the antennal scale so common in the Macrura is present on the lower antenna.") If any one will furnish me with an Amphipod or an Isopod with Nauplius-brood, the existence of which would not be more remarkable in independently produced species than that of a Prawn with Nauplius-brood, I will abandon the whole Darwinian theory.

With regard to the Crabs, and also to the Isopoda and Amphipoda, we were led to the assumption that, about the period when these groups started from the common stem, a simplification of their process of development took place. This also seems to be intelligible from Darwin's theory. When any circumstances favourable to a group of animals caused its wider diffusion and divergence into forms adapting themselves to new and various conditions of existence, this greater variability, which betrays itself in the production of new forms, will also favour the simplification of the development which is almost always advantageous, and moreover, exactly at this period, during adaptation to new circumstances, as has already been indicated with regard to fresh-water animals, this simplification will be doubly beneficial, and therefore, in connexion with this, a doubly strict selection will take place.

So much for the development of the higher Crustacea.

A closer examination of the developmental history of the lower Crustacea is unnecessary after what has been said in general upon the historical significance of the young states, and the application of this which has just been made to the Malacostraca. We may see, without further discussion, how the representation given by Claus of the development of the Copepoda may pass almost word for word as the primitive history of those animals; we may find in the Nauplius-skin of the larvae of Achtheres and in the egg-like larva of Cryptophialus, precisely similar traces of a transition towards direct development, as were presented by the Nauplius-envelope of the embryos of Mysis and the maggot-like larva of Ligia, etc.

It will be sufficient to indicate an essential difference in the process of development in the higher and lower Crustacea. In the latter all new body-segments and limbs which insert themselves between the two terminal regions of the Nauplius, are formed in uninterrupted sequence from before backwards; in the former there is further a new formation in the middle of the body (the middle-body), which pushes itself in between the fore-body and the abdomen in the same way, as these have done on their part between the head and tail of the Nauplius. Thus, that which appears probable even from the comparison of the limbs of the adult animal, finds fresh support in the developmental history, namely, that the lower Crustacea, like the Insects, are entirely destitute of the region of the body corresponding to the middle-body of the Malacostraca. It seems probable that the swimming feet of the Copepoda, as also of the pupae of Cirripedia and Rhizocephala, represent the abdominal feet of the Malacostraca, that is to say, are derived by inheritance from the same source with them.

It would be easy to weave together the separate threads furnished by the young forms of the various Crustacea, into a general picture of the primitive history of this class. Such a picture, drawn with a little skill, and finished in lively colours, would certainly be more attractive than the dry discussions which I have tacked on to the developmental history of these animals. But the mode of weaving in the loose threads would still in many cases be arbitrary, and to be effected with equal justice in various ways; and many gaps would still have to be filled up by means of more or less bold assumptions. Those who have not wandered much in this region of research would then readily believe that they were standing upon firm ground, where mere fancy had thrown an airy bridge; those acquainted with the subject, on the other hand, would soon find out these weak points in the structure, but would then be easily led to regard even what was founded upon well considered facts, as merely floating in the air. To obviate these misconceptions of its true contents from either side, it would be necessary to accompany such a picture throughout with lengthy, dry explanations. This has deterred me from further filling in the outline which I had already sketched.

I will only give, as an example, the probable history of the production of a single group of Crustacea, and indeed of the most abnormal of all, the RHIZOCEPHALA, which in the sexually mature state differ so enormously even from their nearest allies, the Cirripedia, and from their peculiar mode of nourishment stand quite alone in the entire animal kingdom.

I must preface this with a few words upon the homology of the roots of the Rhizocephala, i.e. the tubules which penetrate from its point of adhesion into the body of the host, ramify amongst the viscera of the latter, and terminate in caecal branchlets. In the pupae of the Rhizocephala (Figure 58) the foremost limbs ("prehensile antennae") bear, on each of the two terminal joints, a tongue-like, thin-skinned appendage, in which we may generally observe a few small strongly refractive granules, like those seen in the roots of the adult animal. I have therefore supposed these appendages to be the rudiments of the future roots. A perfectly similar appendage, "a most delicate tube or ribbon," was found by Darwin in free-swimming pupae of Lepas australis on the last joints of the "prehensile antennae." From the perfect accordance in their entire structure shown by the pupae of the Rhizocephala and Cirripedia, there can be no doubt that the appendages of Sacculina and Lepas, which are so like each other and spring from the same spot are homologous structures.

Now in three species of Lepas, in Dichelaspis Warwickii and in Scalpellum Peronii, Darwin saw, on tearing recently-affixed animals from their point or support, that a long narrow band issued from the same point of the antennae; its end was torn away, and in Dichelaspis, judging from its ragged appearance, it had attached itself firmly to the support. From this it follows that this appendage in Lepas australis can hardly be anything but a young cement-duct. If, therefore, the supposition that the appendages on the antennae of the pupae of Rhizocephala are young roots be correct, the roots of the Rhizocephala are homologous with the cement-ducts of the Cirripedia. And this, strange as it may appear at the first glance, seems to me scarcely doubtful. It is true that the act of adhesion of the Rhizocephala has never yet been observed, but it is more than probable that they attach themselves, just like the Cirripedia, by means of the antennae, and that therefore the points of attachment in the two groups indicate homologous parts of the body. From the point of attachment in the Rhizocephala the roots penetrate into the body of the host, whilst in the Cirripedia, the cement-ducts issue from the same point. The roots are blind tubes, ramified in different ways in different species. The cement-ducts in the basis of the Balanidae likewise constitute a generally remarkably complicated system of ramified tubes, with regard to the mode of termination of which nothing certain has yet been made out. Individual caecal branches are not unfrequently seen even in the vicinity of the carina; and, at least in some species, in which the cement-ducts divide into extremely numerous and fine branchlets, forming a network which gradually becomes denser towards the circumference of the basis, these seem nowhere to possess an orifice.

Now as to the question: How were Cirripedia converted by natural selection into Rhizocephala?

A considerable number of existing Cirripedia settle exclusively or chiefly upon living animals;—on Sponges, Corals, Mollusks, Cetaceans, Turtles, Sea-Snakes, Sharks, Crustaceans, Sea Urchins, and even on Acalephs. Dichelaspis Darwinii was found by Filippi in the branchial cavity of Palinurus vulgaris, and I have met with another species of the same genus in the branchial cavity of Lupea diacantha.

The same thing may have taken place in primitive times. The supposition that certain Cirripedes might once upon a time have selected the soft ventral surface of a Crab, Porcellana or Pagurus, for its dwelling-place, has certainly nothing improbable about it. If then the cement-ducts of such a Cirripede instead of merely spreading on the surface, pierced or pushed before them the soft ventral skin and penetrated into the interior of the host, this must have been beneficial to the animal, because it would be thereby more securely attached and protected from being thrown off during the moulting of its host. Variations in this direction were preserved as advantageous.

But as soon as the cement-ducts penetrated into the body-cavity of the host and were bathed by its fluids, an endosmotic interchange must necessarily have been set up between the materials dissolved in these fluids and in the contents of the cement-ducts, and this interchange could not be without influence upon the nourishment of the parasite. The new source of nourishment opened up in this manner was, as constantly flowing, more certain than that offered by the nourishment accidentally whirled into the mouth of the sedentary animal. The individuals favoured in the development of the cement-ducts now converted into nutriferous roots, had more than others the prospect of abundant food, of vigorous growth, and of producing a numerous progeny. With the further development, assisted by natural selection, of the roots embracing the intestine of the host and spreading amongst its hepatic tubes, the introduction of nourishment through the mouth and all the parts implicated in it, such as the whirling cirri, the buccal organs, and the intestine, gradually lost their importance, became aborted by disuse, and finally disappeared without leaving a trace of their existence. Protected by the abdomen of the Crab, or by the shell inhabited by the Pagurus, the parasite also no longer required the calcareous test, in which, no doubt, the first Cirripedes settling upon these Decapods rejoiced. This protective covering, having become superfluous, also disappeared, and there remained at last only a soft sack filled with eggs, without limbs, without mouth or alimentary canal, and nourished, like a plant, by means of roots, which it pushed into the body of its host. The Cirripede had become a Rhizocephalon.

If it be desired to form a notion of what our parasite may have looked like when half way in its progress from the one form to the other, we may consult the figures given by Darwin, (Lepadidae Plate 4 figures 1 to 7) of Anelasma squalicola. This Lepadide, which lives upon Sharks in the North Sea, seems, in fact, to be in the best way to lose its cirri and buccal organs in the same manner. The widely-cleft, shell-less test is supported upon a thick peduncle, which is immersed in the skin of the Shark. The surface of the peduncle is beset with much-ramified, hollow filaments, which "penetrate the Shark's flesh like roots" (Darwin). Darwin looked in vain for cement-glands and cement. It seems to me hardly doubtful, that the ramified hollow filaments are themselves nothing but the cement-ducts converted into nutritive roots, and that it is just in consequence of the development of this new source of nourishment, that the cirri and buccal organs are in the highest degree aborted. All the parts of the mouth are extremely minute; the palpi and exterior maxillae have almost disappeared; the cirri are thick, inarticulate, and destitute of bristles; and the muscles both of the mouth and cirri are without transverse striation. Darwin found the stomach perfectly empty in the animal examined by him.

...

Having reached the Nauplius, the extreme outpost of the class, retiring furthest into the gray mist of primitive time, we naturally look round us to see whether ways may not be descried thence towards other bordering regions. By the structure of the abdomen in Nauplius we might be reminded, like Oscar Schmidt, of the moveable caudal fork of the Rotatoria, which many regard as near allies of the Crustacea, or at any rate of the Arthropoda; in the six feet surrounding the mouth we might imagine an originally radiate structure, and so forth. But I can see nothing certain. Even towards the nearer provinces of the Myriopoda and Arachnida I can find no bridge. For the Insecta alone, the development of the Malacostraca may perhaps present a point of union. Like many Zoeae, the Insecta possess three pairs of limbs serving for the reception of nourishment, and three pairs serving for locomotion; like the Zoeae they have an abdomen without appendages; as in all Zoeae the mandibles in Insects are destitute of palpi. Certainly but little in common, compared with the much which distinguishes these two animal-forms. Nevertheless the supposition that the Insecta had for their common ancestor a Zoea which raised itself into a life on land, may be recommended for further examination.

Much in what has been adduced above may be erroneous, many an interpretation may have failed, and many a fact may not have been placed in its proper light. But in one thing, I hope, I have succeeded,—in convincing UNPREJUDICED readers, that Darwin's theory furnishes the key of intelligibility for the developmental history of the Crustacea, as for so many other facts inexplicable without it. The deficiencies of this attempt, therefore, must not be laid to the charge of the plan drawn out by the sure hand of the master, but solely to the clumsiness of the workman, who did not know how to find the proper place for every portion of his material.